The Western World; Or, Travels in the United States in 1846-47
by Alexander Mackay

Address:
HardPress
8345 NW 66TH ST #2561
MIAMI FL 33166-2626
USA
Email: info@hardpress.net

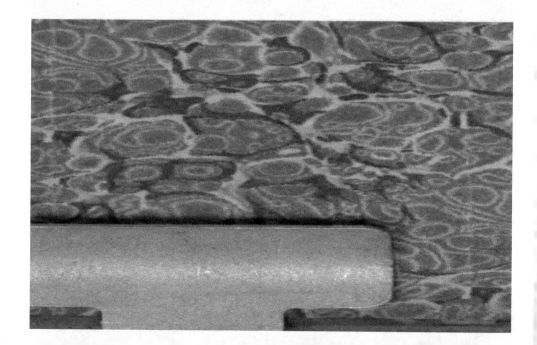

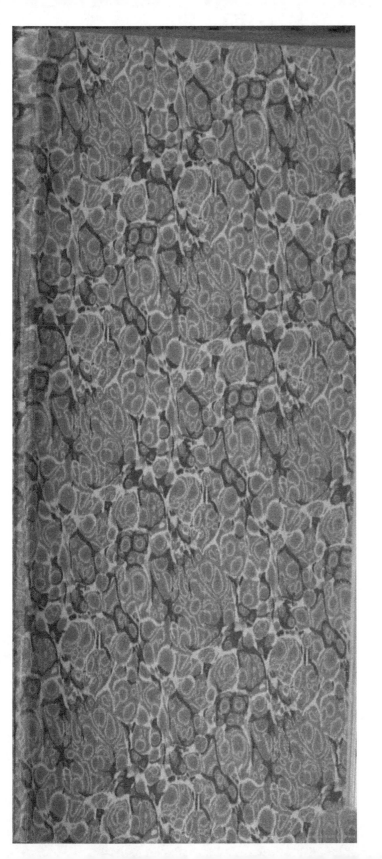

THE

WESTERN WORLD;

OR,

TRAVELS IN THE UNITED STATES

IN 1846-47:

EXHIBITING THEM IN THEIR LATEST DEVELOPMENT,
SOCIAL, POLITICAL, AND INDUSTRIAL;

INCLUDING A CHAPTER ON

CALIFORNIA.

WITH A NEW MAP OF THE UNITED STATES,
SHOWING THEIR RECENT TERRITORIAL ACQUISITIONS, AND
A MAP OF CALIFORNIA.

BY ALEX. MACKAY, ESQ.

OF THE MIDDLE TEMPLE, BARRISTER AT LAW.

IN THREE VOLUMES.
VOL. III.

THIRD EDITION.

LONDON :

RICHARD BENTLEY, NEW BURLINGTON STREET,

𝔓𝔲𝔟𝔩𝔦𝔰𝔥𝔢𝔯 𝔦𝔫 𝔒𝔯𝔡𝔦𝔫𝔞𝔯𝔶 𝔱𝔬 𝔥𝔢𝔯 𝔐𝔞𝔧𝔢𝔰𝔱𝔶.

1850.

LONDON :
R. CLAY, PRINTER, BREAD STREET HILL.

CONTENTS.

CONTENTS.

THE WESTERN WORLD;

OR,

TRAVELS IN THE UNITED STATES IN 1846-7.

———◆———

CHAPTER I.

THE VALLEY OF THE MISSISSIPPI.——FROM NEW ORLEANS TO VICKSBURG.

An unexpected Meeting. — Departure from New Orleans. — The Mississippi—Its Dimensions.—Part which it will yet play in the Drama of Civilization.—Scenery on its Banks.—A Mississippi Steamer.—Fellow-travellers.—Gamblers again.—An Incident.—The State of Mississippi.—Repudiation Case of Mississippi.—The Insolvent States.—The Solvent States.—The Unindebted States.—Responsibility of the States.—Natches.—Vicksburg.—A summary Trial and Execution.—Lynch Law.—Administration of the Law throughout the Union.—Position of the People of the West and South-west.—Allowances which should be made for them.

ON the day previous to that on which, after more than a week's sojourn, I quitted New Orleans, I was delighted, on taking my seat at the *table d'hôte* of the St. Charles, in company with about 500 other guests, to find a valued friend, Mr. D—— from Baltimore, seated next to me on the right. He was an Englishman in the prime of life, but had been so long resident in

America, and had made it the scene of such extensive business operations, that he now combined with an ineradicable affection for his native country a very great partiality for that of his adoption, and with the feelings and sentiments of an Englishman, much that is characteristic of the American. He had never been naturalized, but he was now beginning to reconcile himself to the idea of transferring his allegiance, as he was of becoming a Benedict; his object in contemplating the process of naturalization having less reference to himself than to those who might yet surround him in an endearing relationship. My advice to him was to take no step that he was not certain was necessary; but if he was tired of being sole monarch of himself, to marry first, and wait the tide of events. The process of naturalization was a brief and a sure one when entered upon; the necessity for it in his case had not yet become obvious.

After we had interchanged the ordinary salutations to which such unexpected meetings invariably give rise, I learnt from him that he had arrived in New Orleans but the preceding day, and that the next was that fixed for his departure. He had just taken a run to the South, he said, to " do a bit of business," which, by giving his personal attention to it, he could accomplish more satisfactorily in a single day than by the correspondence of a month. By the time he reached home his journey would have considerably exceeded in length two thousand miles; but he thought nothing of it, having thoroughly contracted the American aptitude for locomotion, and the indifference which the Americans manifest to distances. It was his intention to return, as he had come, by the route over which I had just passed; but as we had both decided on the same time for departure, I deemed it worth while to try if our routes could not be got to coincide. I therefore proposed to him to ascend the Mississippi and the Ohio with me, a course which would not take him much out of his way, as from the latter stream he could reach home by the Baltimore and Ohio railway. He readily consented to the change, at which I was exceedingly rejoiced, both because he was excellent company, and his knowledge of the country and people would be of great advantage to me.

Next morning at an early hour we left New Orleans for St. Louis. Our journey was confined to the Mississippi, which we were to ascend for upwards of 1,200 miles. We were on board a first-class steamer, and as we receded from the town, and before the first curve of the river had hid it from our view, I thought it, as the morning sun shone brightly upon its spires and cupolas, its massive piles of warehouses, its Levee already swarming with busy thousands, and the spars and rigging and multitudinous funnels which lined its semicircular harbour, one of the finest views of the kind I had ever beheld. In itself the southern capital is in every respect a most interesting town. But it has little that is interesting around it, for it stands, as it were, alone in the wilderness, a city without any immediate environs, to attract the stranger, or to recreate its inhabitants.

The Mississippi! It was with indescribable emotions that I first felt myself afloat upon its waters. How often in my schoolboy dreams, and in my waking visions afterwards, had my imagination pictured to itself the lordly stream, rolling with tumultuous current through the boundless region to which it has given its name, and gathering into itself, in its course to the ocean, the tributary waters of almost every latitude in the temperate zone! Here it was then, in its reality, and I, at length, steaming against its tide. I looked upon it with that reverence with which every one must regard a great feature of external nature. The lofty mountain, the illimitable plain, and the seemingly shoreless lake, are all objects which strike the mind with awe. But second to none of them in the sublime emotions which it inspires, is the mighty river; and badly constituted must that mind be, which could contemplate for the first time with a feeling of indifference a stream which, in its resistless flow, passes through so many climes, and traverses so many latitudes, rising amid perpetual snows, and debouching under an almost tropical sun, and draining into itself the surplus waters of about two millions of square miles!

But the grandeur of the Mississippi consists less in the majestic proportions of its physical aspect than in the part which it is yet destined to play in the great drama of civilized life. It was grand, whilst it yet rolled silently and unknown through the unbroken solitudes of the primeval forest—it was grand, when the indomitable but unfortunate Soto first gazed upon its waters, and when they opened to receive at the hands of his disconsolate band, the corpse of its discoverer—and it was grand, when no sound was heard along its course but the scream of the eagle and the war-whoop of the savage—when no smoke curled and wreathed amid the foliage on its banks but such as arose from the wigwam, and when nothing was afloat upon its surface but the canoe and the tree torn from its roots by the flood. But grander will it yet be, ay far grander, when civilization has tracked it from its mouths to its sources ; when industry has converted its sides into a garden, and speckled them with lively towns and glittering cities ; and when busy populations line its shores, and teem along the banks of all its tributaries. Then, and then only, will the Mississippi fulfil its destiny.

Already, with but nine millions of people in the valley, its whole aspect is changed ; the wilderness has been successfully invaded ; the hum of busy industry is heard along its shores ; towns have sprung up, as if by magic, upon its banks ; the combined banner of science and art waves over its waters ; and hundreds of steamers, with a multitude of other craft, are afloat upon its tide. What scene will it present when the present population of the valley is multiplied by ten, and when, serving as a bond of perpetual union, stronger than treaties, protocols, or the other appliances of diplomacy between more than a dozen sovereign and independent commonwealths, it is the common highway, along which will be borne the accumulated products of their united industry to the ocean! Viewed in the double light of what it is and what it is to be, it is marvellous how some can look upon the Mississippi as nothing more than a " muddy ditch." Muddy it undoubtedly is, but that which renders its current so turgid is but the material torn from distant regions, with which it comes laden to construct new territories in more accessible positions. The opaqueness of its volume is thus but one of the means by which is gradually accomplished a great physical phenomenon. Regarded in connexion with the purposes to which it will yet be applied when civilization has risen to full tide around it, the

Mississippi must be equally an object of interest to the Englishman as to the American—for what Englishman can look with indifference upon that which is yet destined to be the principal medium of communication between the great world and the region which is rapidly becoming the chief theatre for Anglo-Saxon enterprise, and will yet witness the greatest triumphs of Anglo-Saxon energy and skill? He takes, then, but a vulgar view of it who treats as merely so much muddy water running through an unpicturesque country, a stream which, ere many more heads are grey, will exercise so important an influence upon the commercial and political relations of the world.

Nowhere has the Mississippi the majesty of appearance presented, throughout most of its course, by the St. Lawrence. At New Orleans it is scarcely a mile in width, expanding somewhat a short distance above the city, and continuing of an average width of a little more than a mile as far up as its confluence with the Missouri. For a long way beyond that point its size diminishes but little, although its depth is not nearly so great as after the junction. Its depth increases as its volume is enhanced by the contributions of one tributary after another, which accounts for the absence of any apparent enlargement of its size for the last fifteen hundred miles of its course, during which it receives most of its great tributary streams. The current flows at the average rate of three miles an hour, and its increasing volume is accommodated by its increasing depth as it proceeds through the soft alluvial deposit in which it has its bed. As it approaches its outlet, the current gradually diminishes, and will continue still further to diminish,

In a moment the young man's eyes flashed fire—and, had he possessed a weapon, a fatal collision might have been the result, for the other was armed.

"A joke, a joke!" cried the rest of the company, "nothing more; no offence meant." And after some further interposition on their part, the storm was strangled in its cradle.

"A joke that came too near the truth, I fancy," said one over his shoulder, in an under-tone to another, as they immediately afterwards separated.

When about two-thirds of the way to Natches, we passed the line dividing, on the east bank of the river, the State of Louisiana from that of Mississippi; the former continuing upon the west bank for nearly two degrees further to the north. On passing the boundary, and having the State of Mississippi on our right, my mind very naturally reverted to a subject with which the name of that State has for some years been most unfavourably identified. My thoughts at length found vent in expression, and I observed to my companion, that it was a matter of astonishment to me that a State possessed of resources like those of Mississippi, could remain for one hour longer than it could avoid it under the stigma which now rested upon its character.

"The subject with which your mind is now occupied," said he, "is one on which there is much misconception abroad. It is misunderstood both through ignorance and prejudice. Some cannot, and others will not, give it an impartial consideration."

"I have heard the same thing more than once advanced during my peregrinations through the country," I replied, "and am inclined to believe that abroad the case is very much prejudged. I

should much like to know the sentiments regarding
it entertained by one occupying a position in the
country so favourable to a proper appreciation of the
subject as is yours."

" I have no objection to giving you my views,"
said Mr. D——, "but I must first stipulate, that
you will carefully discriminate between my endeavours to place the subject in its proper light, and
any approval on my part of the principle or practice
of repudiation. I demand this, not because I think
that you would willingly misconstrue my motives, or
attribute to me principles which every honourable
mind would scorn to entertain; but because our
countrymen, full of preconceived opinions upon the
subject, are but too ready to denounce every effort
at eliciting the real merits of the case, as nothing
short of a direct advocacy of repudiation."

I readily promised to comply, assuring him that
my object was not to confirm any preconceived notion
of my own, but to get at the truth, no matter to what
inference or conclusion it might lead.

" As to the villainy of repudiation," said he,
" naked, absolute, and unequivocal, there can be no
two opinions amongst honourable men."

To such a proposition I could not but assent.

" If any member of this Confederacy," he observed,
continuing, " or any other community, no matter
where situated, were guilty of such, no man who
valued his own reputation could attempt to raise his
voice in its defence."

I acknowledged the risk any one would run in
doing so.

" Now," continued he, " whilst this is the crime
with which some States are directly charged, and in

which the whole Union is more or less involved, in the opinion of so many abroad; it is a crime of which no member of this Confederacy has as yet been guilty, and of which, I trust, no member of it ever will be guilty."

"For my own part," I observed in reply, "I always discriminated widely between the case of Mississippi and that of the other States, which are either wholly, or have been but temporarily insolvent, and have certainly never, even in word or thought, attempted to involve the innocent with the guilty."

"You select," said he, "the case of Mississippi, no doubt, as the worst in the catalogue. So it is; but even Mississippi is not guilty of the enormities with which she stands charged. Repudiation, in its simple acceptation, is the refusal to pay a debt acknowledged to be justly due. Now, as thus construed, even Mississippi has not been guilty of repudiation. The debt which she has refused to pay is a debt which she does not acknowledge to be justly due. If not fraudulently, she insists that it was, at least, illegally contracted, so that she regards it as a debt which she may, but which she is not bound to pay. Whilst this is the real state of the case, she gets credit for cherishing a conviction of the justice of their claims, at the same time that she sets her creditors at defiance. Her language to them is supposed to be this:—'I owe you the money, and can have no possible objection to your claim; but you may whistle for it, for not one farthing of what I justly owe you shall you receive from me.' Whatever community or individual would hold such language to its or his creditors, must have previously sounded the deepest depths of infamy. It is con-

soling to know that even the Mississippians have not
done this, for even they have the grace left to seek
to shelter themselves behind an excuse for their
conduct."

"There is certainly some sense of honour left," I
observed, "in those who care for explaining away or
extenuating their disgraceful conduct, provided the
endeavour to do so be not solely with a view to
escape the punishment which might otherwise attach
to it. The man who tries to excuse himself for the
commission of a wrong, testifies, to some extent, in
favour of what is right. If the Mississippians are
not the graceless and unblushing repudiators which
they are supposed to be, I should like to know the
nature of their excuse, for upon that depends altogether
the extent to which it can palliate their conduct."

"I by no means wish," replied Mr. D——, "to
screen the State of Mississippi from any obloquy
which may justly attach to her in what I have already
said; my sole object has been to show that even she
has not gone the length to which many suppose or
wish to believe that several of the States have gone;
for even she is, in her own eyes, not without excuse
for what she has done. Whether that excuse be
valid or not is another question. It may not be of a
nature to rescue her from all blame, but the very fact
that she tenders one is sufficient to relieve her from
the grosser charge which is so very generally hurled
against her."

"But her excuse?" said I.

"The entire debt of Mississippi," said he, "has
not been repudiated. It is only a portion of it,
though certainly the greater portion, that has been
thus dealt with. Her excuse for refusing to pay

vision, cross the rivulet which the schoolboy can leap, and thread a mazy course amongst gentle undulations, some of which it is cheaper to tunnel than to turn; but here, cities, towns, and the great marts of commerce lie far apart, and to unite them you have to traverse in long straight lines the boundless plain, penetrate the mountain ridges, intersect the interminable forest, span or ferry the mightiest rivers, and cross morass after morass, all of them yet undrained, and some of them undrainable. Taking them as far as they go, there are no works more solid or substantial, or that exhibit themselves as greater triumphs of the skill and perseverance of a people, than the public works of England. But they are on a small scale when compared with those already executed and projected here, and such as are to be yet projected and executed. People measure the greatness of their works by the scale of the occasion for them. Improvements here are on a scale which the people are accustomed to, but a scale which in England would be considered prodigious. The reason is, that in the one case it is necessary to conform to it, whereas in the other it would be unnecessary to adopt it. There are several of the unfinished canals of America, any one of which would make the circuit of some kingdoms. The American is, therefore, condemned to the alternative of making no improvement at all, or of conforming himself in making them to the scale of circumstances. For the last fifteen years a mania for internal improvements has overspread the face of the earth; Mississippi participated in it. She was poor, and the works which she undertook were great and expensive; but their prospective fruits seemed to justify both the effort and the outlay. But her credit was shaken before they were all completed, and some of them are, for the present, absolutely profitless investments. She went greatly beyond her depth ; but so have too many other States, both in the Old World and in the New. If she was too eager to borrow, so were capitalists too eager to lend."

"All this," said I, " may serve as an excuse for her imprudence, but you have not, in my opinion, exonerated her from the substantial charge against her. In pleading an excuse for the repudiation of her debt, she has paid but a lip homage to common decency."

" But even that shows that a sentiment of honesty still remains ; and so long as that lingers in her bosom, there is hope of her redemption."

" That I believe," I observed, " for even if honour fail to induce her to do so, policy and self-interest will yet prompt her to redeem herself; and I have little doubt but that the day will soon come when she will thoroughly repent of her waywardness, and again hold up her head amongst the nations of the world."

Mr. D—— here interrupted our conversation to point out to me the mouth of the Red River, which entered the Mississippi from the west. What we saw was more where the confluence took place than the confluence itself, an island which had been thrown up by the combined action of the two rivers hiding the junction from our view. This great stream, rising amongst the more easterly ridges of the Rocky Mountains, and within what was once the territory of Mexico, and forming, for part of its course, the dividing line between the two republics, flows for about 1,500 miles before it enters the Mississippi, within monarchy, and consequently in fashion. Were she a republic, her present financial state would be imputed to her as the greatest of her crimes. This is the reason why many, who could have done so, have not discriminated between the case of one State and another in the American Union. They eagerly catch at the perversities of one, which they exhibit as a sample of all the rest. It is thus that the public mind in Europe has been misled; and I am sorry to say that literature has, in too many cases, by self-perversion, lent its powerful aid to the deception."

"But this want of discrimination," observed Mr. D——, "is not confined to the case of the insolvent States alone. It is also too much the fashion in England to speak of all the States as if they had, without exception, repudiated their obligations. They forget, or rather will not remember, that, whilst some of the States are free from debt, altogether, the majority of them, being more or less in debt, are solvent, like Great Britain herself, and quite as likely to continue so. But they are all flippantly spoken of, as if, in the first place, they were one and all insolvent; and, in the next, had one and all repudiated their debts."

"There is much truth," said I, "in what you urge, and I must confess that nothing can be more unfair."

"But the most extraordinary thing connected with this whole matter," said he, "is the call which is made by some who are ignorant of the relationship in which the different States stand towards each other, and by others who thoroughly understand it, upon the solvent States, to pay, or to aid in paying, the debts of such as are in default. What encouragement would a man have to pay his own way in the

world, if he were liable to be called upon to clear
the scores of his neighbours ?   Of what avail would
it be to New York to keep herself out of debt, or,
in contracting obligations, to respect the limits of
her solvency, if she were liable to be involved in the
extravagances which might be committed by any or
by all of the neighbouring communities ?"

"But this call," I observed, interrupting him,
"upon the solvent States to assume, in part, the
debts of their confederates, is based upon the supposition that they are each but a component part of one
great country."

"And so they are," replied he, "for certain purposes, but not for all.   A, B and C unite in copartnership, for the avowed purpose of manufacturing
certain kinds of goods, but for none other.   If the objects of the copartnership are published to the world,
it would be unreasonable to hold that they were
bound together for purposes not specified amongst
these objects.   In any transaction connected with the
business of the firm, any one of the partners can bind
all the rest.   But in transactions notoriously alien to
the business of the firm, it is not competent for any
one partner to bind his fellows ; and any one giving him
credit in such transactions, does so upon his own sole
responsibility.   Should the security of the individual
fail in such a case, the creditor would be laughed at
who would call upon the firm to liquidate the debt.
And so it is with the Federal Union.   The States of
which it is composed are bound together in a political
relationship, for certain specified objects, and for
none but such as are specified.   To carry these out,
certain powers are conferred upon them in their
federal and partnership capacity.   The power to

borrow money for local purposes is not one of these ; and as one State has no power to borrow money for another, nor all the States together for one State, there is but little justice in calling upon one State to pay the debts of another, or on all the States to pay the debts of any one or more which may be in default. There is this difference, too, between the Union and a common partnership, that whereas in the latter one member of the firm can bind all, provided the transaction be within the objects of the partnership ; in the former, it is competent to no one State to bind the rest, even in matters common to all the States, and within the purview of the objects for which they are united. In such case it is the general government alone that can be dealt with, as the sole agent and representative of the Union. If any one gives credit to it, the Union, that is to say all the States, are responsible ; but when credit is extended to a particular State, it is to that State alone that the creditor can justly look for his reimbursement."

" I am aware," said I, " that the objects for the accomplishment of which the money was borrowed, were matters within the exclusive control of the indebted States themselves ; and that, therefore, the credit could only have been given exclusively to them. But you must admit that the line of demarcation between local and federal powers, and local and federal responsibility, is not very generally understood in Europe."

" But," replied he, " there is no reason why the inhabitants of Delaware, which owes nothing, or of New York, which pays what it owes, should pay the penalty of the ignorance, real or assumed, of the money-lenders in Europe, who chose to deal, without their knowledge, or without getting their security, with the State of Mississippi. The terms and conditions of the federal compact are no secret. They have been patent to the world for the last sixty years. What more could be done to give them publicity than has been done? When a State goes into the money-market to borrow, she does not do so under the shelter of a secret or ambiguous deed of copartnership, by which the money-lender may be deceived, but as a member of a confederacy, bound together by a well-known instrument, which notoriously confers no power upon her in borrowing money to pledge the credit of any of her confederates. The States of Germany are knit together in one federal union for certain purposes, but their common responsibilities terminate when the limits of these purposes are reached. The borrowing of money for local purposes is not one of the objects of the German Confederation. Would it be competent, then, for an English capitalist who had lent money to Saxony, which she omitted to return him, to call upon Austria or Bavaria to make good his loss? And the same with the American Union. The powers and responsibilities of the States are, or should be, as well known to the capitalist as those of the States of the German Confederation. And in truth, there is reason to believe that they were well known when the money was advanced, and that the plea of ignorance is a sham plea, preferred more to move the sympathies than to appeal to the justice of the other States. He who lent, then, to Mississippi or Illinois, on the sole responsibility of Mississippi or Illinois, has obviously no claim in law, or in equity, against any State but Mississippi or Illinois. If he lent on what he considered at the time a doubtful security, in the hope that, should that security fail him, the other States, which had no knowledge of, or benefit from, the transaction, would either be moved by compassion to save him harmless, or shamed by a false cry into so doing, his conduct was not such as would bear the test of a rigid scrutiny. Such a course is as questionable as lending to a man of doubtful credit, on the speculative security of his numerous friends."

" On this point," I observed, " I can find no flaw in the argument which you advance. It is obvious that, when a man lends money upon a particular security, he cannot afterwards look for its repayment to parties whom he could not have legally or morally contemplated as involved in the benefits or responsibilities of the transaction at the time of its occurrence. Besides, if one State was liable for the debts of another, it should have some control over the expenditure of the other. And when we consider that one State borrows money for the construction of works, which, when in operation, will injuriously affect similar works in another, it would be especially hard were that other to be held answerable for its default. And so with the general Government. It has no control over local expenditure; and it would be monstrous, therefore, to make it responsible for local liabilities. But if I mistake not, the project of the general Government assuming the State debts has found much favour even in this country."

" It has," replied he, " though not as a matter of right, but simply as one of expediency. The general credit was affected by the misconduct of a few of the members of the Union, and to rescue all from an odium that justly attached but to the few, the propo-

sition you allude to was made.   But the proposition
was, not so much in its principle, as in its incidents,
one to which the solvent and unindebted States could
not agree ; the consideration for which the assump-
tion was to be made being one in which they were
as much interested as the insolvent States themselves.
They could not, therefore, consent to a proposal which
would have virtually taxed them to pay a portion
of the debt of the delinquents.   It has thus, for the
present, been abandoned, and it is to be hoped that
ere it is again mooted, the defaulting States will be
restored to solvency."

Our conversation, which embraced the whole subject, and of which this is but an epitome, was here
interrupted by our approach to Natches.   My mind
continued for some time to dwell upon the subject,
which the more I learnt regarding it, I was the
more convinced was misunderstood.   To involve the
whole Confederacy in the crimes or misfortunes of a
few of its members is obviously unjust.   It is but fair
that a wide discrimination should be made between
the guilty and the innocent.   This can only be done
by taking the States separately, and dealing out our
judgments in regard to each, according to the position in which we find it.   And, in applying this
rule, let us bear in mind that they are divisible into
four classes.   In the first, Mississippi is alone comprehended ; for she alone has repudiated, although
she has not been so graceless as to do so without all
excuse.   The second comprehends the few States
whose treasuries have been bankrupt, but none of
which have ever repudiated their obligations.   Some
of these have resumed payment, and are once more
in a state of perfect solvency.   In the third class
are embraced the majority of the States, and such

as have ever been solvent, neither repudiating the claims against them, nor omitting to pay them. The fourth class comprises the few States which are so fortunate as to be entirely free from public debt. And when the European talks of the American people doing justice to the public creditor—meaning thereby that the whole Union should saddle itself with the debts of a few of its members, contracted with the knowledge of their creditors, upon their own sole responsibility—he should remember that there is justice also on the other side, and that the people of Delaware and North Carolina, who owe nothing, and those of New York and other States who are paying what they do owe, cannot, with any degree of propriety, be called upon to bear the burden of transactions entered into by others for their sole benefit, and to which they alone were parties. There is but little either of morality or justice in seeking to involve parties in the responsibility of transactions with which they have had nothing whatever to do.

And in dispensing blame to the parties really deserving it, it is not always to the inculpated States that we are to confine our censure. What injured them was precipitate speculation. This is promoted as much by the capitalist as by the borrower, and in many cases more so. The time was when nothing but a foreign investment would satisfy the English capitalist. A home or a colonial speculation stunk in his nostrils; nothing but that which was foreign would satisfy him. The foreigner seeing an open hand with a full purse in it extended to him, was tempted to grasp at it, and his appetite for speculation was quickened by the ease with which he obtained the means of pandering to it.   At this very time our magnificent colonies in North America were demanding accommodation, but could not procure it.   The six per cent. which they modestly offered, was refused for the seven, eight, and ten per cent. offered by the neighbouring States, which by the very favouritism thus shown them were encouraged to endeavour to outrun each other in their mad career.   They are truly to be pitied who, having had no hand in the original transactions, are now the innocent holders of the bonds which have been repudiated, or which remain unpaid. But they can only justly look for their indemnity to the security on which they were contented to rely, without seeking to involve others in their misfortunes who are as innocent as themselves.

Credit has been described as a plant of tender growth, which the slightest breath may shrivel.   There is no doubt but that the conduct of some of its members occasioned a severe shock to the credit of the whole Union.   For a time all the States were treated as if, without exception, they had been involved in a common delinquency.   But this injustice did not last long, and the solvent States are being gradually reinstated in their former credit and position.   And even now, as permanent investments, many, and not without reason, regard American securities as preferable to all others.   The credit of the general Government is at present much more in vogue than that of any of the States; but as permanent investments, the securities of the States are to be preferred to those of the general Government.   Should the Union fall to pieces, the general Government will be extinguished in the crash, but the States will preserve their identity whatever may become of the Confederation. And notwithstanding the stigma which for some time has unfortunately attached to her name, there is no State in the Union which can offer greater inducements to permanent investment than Pennsylvania. Her resources are greater and more varied than those of any of her confederates, and her future wealth will depend upon their development. What these resources are in their extent and their variety, and how far her position is such as will necessarily call them into speedy and active requisition, will be inquired into in a subsequent chapter.

At Natches, which is one of the largest and most prosperous towns in the State, and situated mainly on a high bluff overlooking the river, we remained but a sufficient time to land and to receive passengers, and to take in a fresh supply of fuel and provisions. We had already stopped at several road-side stations, as they might be called, for the purpose of replenishing our stock of wood, the quantity consumed by the furnaces being enormous. From Natches we proceeded towards Vicksburg, also in the State of Mississippi, and about 106 miles higher up the river.

The name of this place suggested at once to my mind a terrible incident, of which some years ago it was the scene, and which strongly illustrates a very unfavourable feature of American life in the South-west. The gamblers and blacklegs, who had made Natches too hot to hold them, made the town of Vicksburg their head quarters, and as they increased in numbers, so increased in boldness, and carried matters with so high a hand, as for a time to terrify and overawe the more honestly disposed of their fellow-citizens. The evil at length attained a magnitude which determined the better portion of the inhabitants at all hazards to put it down ; and as the law was too weak to reach the ruffians, it being as difficult to obtain a conviction against them as it is to get one against a repealer in Ireland, a summary process of dealing with them was resolved upon.   A number of them were accordingly surprised when engaged in their nefarious practices, some of whom escaped in the confusion, leaving about half a dozen in custody.   These were conveyed a short distance out of the town, and after a summary trial and conviction by Lynch law, were hanged upon the adjacent trees.   Lawless and horrible as this act undoubtedly was, the terrible vengeance which it inflicted upon a set of blackguards, who harassed and systematically annoyed the community, had a salutary effect for a time ; the survivors, if they did not abandon their practices, paying a little more respect to public opinion in their mode of pursuing them.   The effects of the lesson then administered, however, have by this time pretty well worn off, if one may judge from the numbers in which the southern portion of the Mississippi and its tributaries are yet infested by the vagabonds in question, and the openness with which they are beginning again to prosecute their iniquitous vocation.

The excesses thus occasionally committed by the populace in the South under the designation of Lynch law, are much to be deplored, although they are almost necessarily incident to a state of society in which public opinion is yet weak and but equivocally pronounced—in which the law is feebly administered, and which exists in the midst of circumstances less favourable than those by which we are surrounded for the enforcement of public morality, and the due administration of justice. To those conversant with the real condition of society in the South-west, the wonder is not so much that Lynch law has been so frequently resorted to, as that the ordinary law has not been more frequently departed from. The population of the immense areas which bound the Southern Mississippi on either side is but yet scanty, people in general living far apart from each other. Add to this that the war which they are carrying on, each in his comparatively isolated position, against nature, has a tendency more or less to bring the civilized man in his habits, tastes, and impulses, nearer to the savage, and to impart asperities to the character which are rubbed off by an every day contact with society. No position that is not actually one of barbarism, could be more favourable than that of the western pioneer to the inculcation of the law of might, his life being not only a constant warfare with the wilderness, but his safety, from the nature of the dangers with which he is surrounded, chiefly depending upon his own vigilance and presence of mind. He is thus daily taught the habit of self-reliance, instead of looking to society for his security. It is scarcely to be wondered at that men so circumstanced, and, as it were, so educated, should occasionally take the law into their own hands, instead of resorting for justice to tribunals far apart from them, to reach and attend which would be accompanied by great loss of time and money, and which might after all fail in rendering them justice.

In the Northern and Eastern States the law is as regularly administered as it is in England, and life and property are as safe under its protection as they

are in any country within the pale of civilization.
But most of these States have been long settled, the
wilderness in them has been reduced, society has become dense, and exists in the midst of all the appliances of civilization; its members can rely upon each
other for support in carrying out the law, and they
prefer the security of society to any that they could
attain for themselves; and, which is very important,
their tribunals are numerous, respectable, and near at
hand.   From a people so situated we are quite right
in exacting a strict conformity to the practices of
civilized life.   But when we go further, and exact the
same of the people in the extreme West and South-
west, we either forget that they are differently circumstanced, or deny that circumstances have any influence on social and individual life.   Transplant to
the regions beyond the Mississippi a colony of the
most polished people, either from Old or New England, and let them be circumstanced precisely as the
western pioneers are, and how long would they retain
their polish, or be characterised by those amenities,
or exercise that mutual reliance upon each other,
which marked their life and habits in their former
abode?   Bring the polished man in contact with
savage nature, which he is called upon daily to
subdue, that he may obtain his daily bread, and the
one must succumb to the other, or both will undergo
a change.   As man civilizes the wilderness, the wilderness more or less brutalizes him.   In thus elevating nature he degrades himself.   And thus it is with
the pioneers of civilization in the American wilds.
Generally speaking, they have not had the advantage
of a previous polish.   Born and brought up in the
midst of the wilderness, they fly rather than court

the approach of civilization. They care little for the open fields which their own labour has redeemed; they love the recesses of the forest, and regularly retire before it as population advances upon them. This hardy belt of pioneers is like the rough bark which covers and protects the wood, and serves as a shield under shelter of which the less hardy and adventurous portions of the community encroach upon the wilderness. To expect them rigidly to conform to all the maxims of civilized life would be but to expect civilization to flourish in the lap of barbarism. Even yet, along the borders of conterminous countries which we call civilized, how often do we find lawlessness and violence prevailing to a deplorable extent! And is our sense of propriety to be so greatly shocked when we find them occasionally manifesting themselves upon the American border, where the domain of civilization is conterminous with that of the savage, the buffalo, and the bear? Every excess committed in these remote, wild, and thinly peopled regions is to be discountenanced and deplored; but we should not visit them with that severity of judgment which such conduct amongst ourselves would entail upon those who were guilty of it. As the wilderness disappears, and the country becomes cultivated, the civilization of nature will react beneficially upon those, or the descendants of those, who were instrumental in rescuing her from the barbarism in which she was shrouded; population will become denser and more refined, and man will rely more upon his social than his individual resources. When this occurs, and the portion of the country now considered is thus brought within the pale of civilization, we may exact, and exact with justice, from its people a strict amenability to all the requirements of civilized life. But before it occurs we should not overlook their circumstances in dealing with their conduct. Even in the most civilized communities departures are sometimes deemed necessary from the ordinary principles by which society is regulated, and from the ordinary safeguards by which it is secured. We need not be surprised if exceptions to general principles occur where society is as yet but in a state of formation; and it may be that, in the semi-civilized regions of America, the dread tribunal of Judge Lynch may sometimes be as necessary, as, in civilized life, are states of siege, and the supersession of the ordinary tribunals of justice by martial law.

A better order of things is now making its appearance along the banks of the lower Mississippi, where public opinion is fast gaining ground upon the lawless disturbers of the public peace. In some cases the carrying of arms is now forbidden—a most prudent measure, as it frequently happens that to be prepared for war is the very worst guarantee for peace. Society is gradually feeling its strength, and once convinced of it, will know how to take measures for its own security. The first and worst epoch in its history is past. It has survived a perilous infancy, and is now advancing to maturity; and the moral aberrations of which, in its youth, it may have been guilty, may yet be to it as the complicated diseases of childhood are to the boy, in preparing him for becoming the healthy man.

CHAPTER II.

THE VALLEY OF MISSISSIPPI. — AGRICULTURE AND AGRICULTURAL INTEREST OF THE UNITED STATES.

Vicksburg.—The Walnut Hills.—The Arkansas and the Tennessee.
—Variety of Craft met with upon the River.—Difference between
the two Banks.—Memphis.—Posthumous adventures of Picayune
Walker.—Conversations on Slavery.—A Race.—Days and Nights
on the River.—The Mouth of the Ohio.—Change of Scene.—St.
Louis.—Who are the Yankees?—Description of St. Louis.—Its
Commercial Advantages and Prospects.—The American Prairie.
—Agriculture and Agricultural Interest of America.—Five great
Classes of Productions.—Five great Regions corresponding to them.
—The Pasturing Region.—The Wheat and Tobacco growing Re-
gions.—The Cotton and Sugar Regions.—Cost at which Wheat
can be raised on Prairie land.—The surplus Agricultural Pro-
ducts of America.

ON leaving Vicksburg, which is charmingly situated
on a high sloping bank, formed by the bluffs into
what appears to be a series of natural terraces, which
render it much more accessible than Natches, we
steamed rapidly up the river, having as yet, although
about four hundred miles from New Orleans, accomplished but one-third of our journey. The Walnut
hills, which come rolling down to the water's edge
immediately above Vicksburg, are exceedingly picturesque, mantled as they are to the top in a rich
covering of grass and foliage. Beyond them the
right bank sinks again, and presents to the eye, for

many miles, an unbroken succession of extensive and flourishing cotton plantations.

We soon left the State of Mississippi behind us, and had that of Tennessee on our right, and for some distance Arkansas on our left. Both these States are named from the chief rivers flowing through them to swell the volume of the Mississippi, the Arkansas directly, the Tennessee indirectly, by uniting with the Ohio. Both streams are upwards of 1,200 miles long, and navigable to steamers for hundreds of miles. The Arkansas, like the Red River, rises in the Rocky Mountains, and after flowing in a south-easterly direction through the State to which it gives its name, enters the Mississippi on its west bank. We had already passed the junction on our left, as we had also the mouths of several other rivers entering on the same side, which in Europe would be considered first-class streams.

It is almost impossible to describe the variety of craft which we met upon the river. We passed and saluted steamers innumerable, generally crowded with passengers. Others were so overloaded with cotton-bales, as to present more the appearance and proportions of a long hay-rick than of any other known terrestrial object. There were flat boats innumerable, precipitous at the sides, and quite square at either end, sometimes with an apology for a sail hoisted upon them, and sometimes with an oar out on either side to help them to drop down with the strong heavy current. It is not many years since this was the only craft known on the Mississippi, being constructed with sufficient strength to bear the voyage down, for they never attempt the re-ascent of the stream. When they have served their purpose, on reaching their destination they are broken up, and the materials disposed of to the best advantage. Before the introduction of steamers, travellers had to ascend into the interior by land. Then again we would meet a family emigrating from one part of the valley to another, by dropping down in a rudely constructed barge, which would yet be broken up and converted into a hut or "shanty" on shore. There were floating cabins too, which would only have to be dragged ashore on reaching their destination. And then came floating "stores," containing calicos, cloths, pots, pans, groceries and household wares of all descriptions,—the pedlars of these regions very wisely conforming themselves to the nature of their great highways. And instead of caravans, as with us, upon wheels, there were shows and exhibitions of all kinds afloat, in some of which Macbeth was performed, Duncan being got rid of by throwing him into the river instead of stabbing him. Here and there too, a solitary canoe and small boat would cross our track, as would also occasionally a raft, some of the timber constituting which may have been purchased, but all of which the raftsmen undoubtedly intended to sell. In short, it was a source of amusement to us to watch the varied and generally unshapely contrivances in the way of craft, many of them laden with live stock, which the "Father of waters" bore upon his bosom.

All the way from Baton Rouge, in Louisiana, the scenery on our right was more or less varied by gentle undulations, sometimes attaining the dignity of hills ; whilst the river, with occasional gaps, some of which extended for many miles, was lined by a succession of bluffs, whose different heights and forms gave constant novelty to the scene. In some places they rose over the water for several hundreds of feet, a low ledge of land generally intervening, where they were highest, between them and the river. It is on these ledges that the lower portions of the chief towns on this bank are built. The cliffs, when the water is in direct contact with them, are soon worn away beneath, when the superincumbent mass gives way, forming the ledges in question. These again are in time washed away by the river, when the cliffs are again attacked, and with the same result. The cliffs continue, with more or less interruption, nearly the whole of the way up to the Ohio. Being generally formed of clay or sand, they are in some places washed by the rains and moulded by the winds into the most fantastic forms, sometimes resembling feudal castles pitched upon inaccessible rocks, and at others being as irregular and grotesque as a splintered iceberg. Very different is the character of the other bank. The whole way from New Orleans up to the mouth of the Ohio, it is, with a few trifling exceptions, one unbroken, unmitigated and monotonous flat. On both sides the land is extremely rich, the cane brake and cypress swamp, however, being frequent features on the west. There are many " Edens " on this side of the river; but the general character of the soil upon it, from the delta to its sources, is of the most fertile description, the spots unfit for human habitation being rare exceptions to the rule. At regular distances are wood stations, on projecting points of land, the wood being obtained from the forest behind, cut upon the spot by negroes, and corded and ready to be taken on board as fuel by the steamers as they pass. The river on this side being

in contact with the very soil, which is soft and alluvial, its greatest encroachments are made upon this
bank. You sometimes pass groves of trees which a
few years ago had stood inland, with their roots now
half exposed, and themselves ready to fall into the
water, some to drift out to sea, and others to become
snags, and render perilous the navigation of the river.
Now and then, too, you make up with groups of
cypresses and palmettos, festooned with Spanish moss;
and sometimes with clumps of the Pride of China,
with wild vines clinging to their trunks and branches.
Here and there also you see, overhanging the stream,
the wreck of what was once a noble forest tree, now
leafless and barkless, holding out its stiff and naked
arms ghastlily in the sun, telling a mournful tale to
the passer-by—the blanched and repulsive skeleton of
that which was once a graceful form of life. Were
the east bank similar to the west, the Mississippi
would, in a scenic point of view, be to the traveller
dreary enough.

As you ascend it you still find the river pursuing
the same serpentine course as below. The bends are
not so great, but quite as consecutive, it being seldom
that the stream is found pursuing a straight course
for many miles together. We could discern on either
side, as we proceeded, many traces of deserted channels; and some of these are to be seen in parts from
thirty to forty miles from the present course of the
stream.

As we approached the town of Memphis in the
State of Tennessee, the bluffs on the right became
more consecutive, loftier, and more imposing in their
effect. Near the town they are in parts almost as
continuous as, though higher and of a darker colour

than, the cliffs in the neighbourhood of Ramsgate; whilst roads are here and there cut through them down to the water's edge, like the deep artificial gullies which are so numerous along the Foreland. Memphis is situated on the top of a very high bluff, so that part of the town only can be seen from the river. There is a small group of houses below the cliff at the landing-place, where several steamers were lying as we approached. In addition to this Memphis in Tennessee, and that which is or was in Egypt, there is another Memphis in Mississippi, *à propos* to which I overheard in New Orleans the following story told by one negro to another:—

" You come from Miss'sippi, don't you, Ginger ? " said the narrator, who was a fine negro and had been in the North.

" To be sure I do, Sam," said Ginger.

" I tell you what it is then, you have no chance no how comin' from that State."

" What are you drivin' at ? " asked Ginger.

" Isn't that the repoodiatin' State ? " demanded Sam.

" To be sure," said Ginger, " but it was'nt the coloured folks, it was the white men did it."

" Well, you may have a chance if you die in Loosiany, but don't die in Miss'sippi if you can help it," said Sam in a confidential tone.

" I won't die no where if I can help it," was Ginger's response.

" Did you know Picayune Walker, who lived to Memphis ? " asked Sam.

" Know'd him well," said Ginger, " but him dead now."

" Well," said Sam, " I was to Cincinnati when he died. De Sunday after I went to meetin'. De color'd gemman who was preachin' tell us that Picayune Walker, when he die, went up to heaben and ask Peter to let him in. 'Who's dat knockin' at de door?' said Peter. 'It's me, to be sure; don't you know a gemman when you see him?' said Picky. 'How should I know you?' said Peter, 'what's your name?' 'Picayune Walker,' he said. 'Well Massa Walker, what you want?' Peter then ask. 'I want to get in, to be sure,' said Picky. 'Where you from, Massa Walker?' den ask Peter. 'From Memphis,' said Picky. 'In Tennessee?' ask Peter. 'No, Memphis Miss'sippi,' said Massa Walker. 'O, den you may come in,' said Peter, a openin' o' de dore; 'you'll be somethin' new for 'em to look at, it's so long since any one 's been here from Miss'sippi.'"

"Him berry lucky for a white man from dat 'ere State," was Ginger's only remark.

On leaving Memphis, I had a long conversation with a southerner on board, on the subject of slavery. Nothing can be more erroneous than the opinion entertained and promulgated by many, that this is a forbidden topic of conversation in the South. I never had the least hesitation in expressing myself freely on the subject in any of the Southern States, whenever an opportunity offered of adverting to it; nor did I find the southerners generally anxious to elude it. Much depends upon the mode in which it is introduced and treated. There are some so garrulous that they must constantly be referring to it, and in a manner offensive to the feelings of those to whom it is introduced. It cannot be denied but that the suicidal and over-zealous conduct of the abolitionists has made the Southerners somewhat sensitive upon the subject; and they are not very likely to listen with complacency to one who, in discussing it, manifests the spirit and intentions of a propagandist. But if calmly and temperately dealt with, there are few in the South who will shrink from the discussion of it; and you find, when it is the topic of discourse, that the only point at issue between you is as to the means of its eradication.

Having strolled with Mr. D—— towards the prow of the boat, I found myself close to where some negroes were busily at work attending to the furnaces. Having replenished them, they set themselves down upon the huge blocks of wood which constituted their fuel, and rubbed the perspiration off their faces, which were shining with it as if they had been steeped in oil.

"See de preacher dat come aboard when we were a woodin' up at Memphis?" asked one named Jim of another who answered to the imperial name of Cæsar.

Cæsar replied in the affirmative, pouting his huge lips, and demanding of Jim to know if he thought that he Cæsar was blind.

"He just marry a rich wife to Memphis, de lady wid him," said Jim, disregarding the interrogatory.

"Dey all do de same," observed Cæsar. "Dey keep a preachin' to oders not to mind de flesh pots, but it's only to grab de easier at dem demselves."

"Pile on de wood, Jim," continued Cæsar, noticing that the furnaces were once more getting low. In a few seconds their ponderous iron doors were again closed, and they blazed and roared and crackled over the fresh fuel with which they were supplied.

"What you sayin' about Massa Franklin few minutes ago?" asked Jim as soon as they were again seated.

" Dat he took fire from heaben," replied Cæsar.

" From de oder place more like," said Jim in a tone of ignorant incredulity.

Cæsar thereupon rolled his eyes about for a few seconds, and looked the caricature of offended dignity. " Will you never larn nothin'?" said he at last, regarding his companion with contemptuous pity.

" Well, how did he do it?" asked Jim.

" Wid a kite to be sure," said Cæsar, getting very unnecessarily into a passion. Jim still looked provokingly incredulous. " I tell you, wid a kite," continued Cæsar, hoping to make himself more intelligible by repetition.

" But how wid a kite?" asked Jim, making bold to put the query.

" Don't you see yet?" said Cæsar; " he tied a locofoco match to it afore he sent it up, to be sure."

" Ah!" ejaculated Jim, getting new light upon the subject, " and lighted it at de sun, didn't he?"

" He couldn't get at de sun, for I told you afore it was cloudy, didn't I?" observed Cæsar.

" Well den, how light de match?" asked Jim, fairly puzzled.

" De cloud rub agin it," said Cæsar, with the air of one conscious of imparting to another a great secret. But his equanimity was again disturbed by the painful thought of his companion's obtusity, and when he called upon him once more to " pile on de wood," it was in connexion with a friendly intimation to him that he was " only fit to be a brack man."

At this moment an ejaculation of " Mind your fires there!" proceeded from the captain, who had approached, and was now standing on the promenade deck between the funnels, and looking anxiously

forward at some object in advance of us. On turning
to ascertain what it was, I perceived a steamer which
had left Memphis on its way up to Louisville about
ten minutes before we did. She was going at half
speed when I first observed her, but immediately put
all steam on. I at once divined what was to take
place. The firemen seemed instinctively to understand it, as they immediately redoubled their efforts to
cram the furnaces with fuel. By the time we were
abreast of the "Lafayette," for that was our rival's
name, she had regained her full headway, and the
race commenced with as fair a start as could well be
obtained. Notwithstanding the known dangers of
such rivalry, the passengers on both boats crowded
eagerly to the quarter-deck to witness the progress of
the race, each group cheering as its own boat seemed
to be leading the other by ever so little. By this
time the negroes became almost frantic in their
efforts to generate the steam ; so much so that at one
time I thought that from throwing wood into the
furnaces, they would have taken to throwing in one
another. But a short time before upwards of two
hundred human beings had been hurried into eternity
by the explosion of a boiler ; but the fearful incident
seemed for the moment to be forgotten, or its warn-
ings to be disregarded, in the eagerness with which
passengers and crew pressed forward to witness the
race. I must confess I yielded to the infection, and
was as anxious a spectator of the contest as any on
board. There were a few timid elderly gentlemen
and ladies who kept aloof ; but with this exception,
the captain of each boat had the moral strength of
his cargo with him. For many minutes the two
vessels kept neck and neck, and so close to each

other, that an explosion on board either would have calamitously affected the other. At length, and when there still appeared to be no probability of a speedy decision, I perceived a reaction commencing amongst those around me, and on the name of the " Helen McGregor " and the " Moselle," two ill-fated boats, being whispered amongst them, many retired to the stern, as far from the boilers as they could, whilst others began to remonstrate, and even to menace.

" How can I give in ? " asked the captain, in a tone of vexation.

" Run him on that 'ere snag, and be d——d to him," suggested the mate, who was standing by.

The snag was about two hundred yards ahead, just showing his black crest above the water. It was the trunk of a huge tree, the roots of which had sunk and taken hold of the soil at the bottom ; about eight inches of the trunk, which lay in a direction slanting with the current, projecting above the surface. From the position which they thus assume snags are more dangerous to steamers ascending than to those descending the current. In the latter case, they may press them under and glide safely over them ; but in the former, the chances are, if they strike, that they will be perforated by them, and sunk. They are the chief sources of danger in navigating the Mississippi. The captain immediately took the hint, and so shaped his course as to oblige the rival boat to sheer off a little to the right. This brought her in a direct line with the snag, to avoid which she had to make a sharp, though a short detour. It sufficed, however, to decide the race, the " Niobe " immediately gaining on the " Lafayette " by more than a length. The latter, thus fairly jockeyed out of her object, gave up the contest and dropped astern. There are certainly laws against this species of racing ; but the Mississippi runs through so many jurisdictions that it is not easy to put them in force. Besides, it was evident to me, from what I then saw, that, in most cases, passengers and crew are equally *participes criminis.*

. We had now been upwards of three days and three nights upon the river, which had varied but little in width, apparent volume, or general appearance, since we first made the bluffs at Baton Rouge. It was curious to awake every morning upon a scene resembling in. everything but a few of its minute details that on which you had closed your eyes the previous night, and with a consciousness that you were still afloat upon the same stream ; and that, whilst asleep, you had not been at rest, but steaming the entire night against the current, at the rate of from eight to ten miles per hour.

Towards the close of the fifth day we were coasting the low shore of Kentucky on our right, with the State of Missouri on our left ; and early on the morning of the sixth, were off the mouth of the Ohio. As we crossed the spacious embouchure, there was one steamer from St. Louis, turning into the Ohio, to ascend it to Pittsburg, 900 miles up ; and another, which had descended it from Cincinnati, just leaving it, and heading down the Mississippi for New Orleans, one thousand miles below. No incident could have occurred better fitted to impress the mind with the vastness of these great natural highways, and their utility to the enormous region which they fertilize and irrigate. The Ohio enters the Mississippi on its east bank, between the States of Kentucky and Illinois, and about 1,100 miles from its mouth.

ever, on the west banks of the Mississippi that prairies most abound, particularly in the States of Arkansas, Missouri, and Iowa. North of the Ohio they are also to be met with in great numbers and of vast extent, the prairies of Illinois being equal in grandeur and extent to any on the opposite side, with the exception, perhaps, of some of those on the Missouri River, some hundreds of miles above its junction with the Mississippi. Those in the neighbourhood of St. Louis, although not remarkable for their extent, give a good idea of them all. In some cases they seem boundless as the ocean, nothing being visible to break the monotonous surface of long waving grass with which they are covered to the very horizon. They are generally interspersed, however, with woodland or solitary clumps of trees, which, particularly where the surface is broken and undulating, as is the case in the country directly north of the Missouri, give them a very picturesque aspect. When the wind sweeps over them the effect is magnificent; the grass bending beneath its tread and undulating like the waves of the green sea. Though not in all cases, they are frequently covered during the summer with wild flowers, successive generations of which, for several months, enamel their surface; some of these flowers being small and modest, and others, the great majority, large, flaunting, and arrayed in the most gorgeous tints. But like the brilliantly plumaged birds of America, which have no song in them, these gaudy prairie flowers have seldom any perfume. I can conceive no greater treat to the florist than to find himself by the margin of an American prairie when thus attired in the gayest robes of summer. They are cleared by burning the grass upon them when it becomes withered and dry. When the fire thus created spreads over a large surface, the effect at night is grand in the extreme. When the wind is high the flames spread with fearful rapidity, rather against than with it, fuel being most plentifully provided for them in this direction by the long grass being bent over the fire. These fires are frequently accidental, and sometimes do great damage to settlers. Instances have occurred in which trapping parties have had the utmost difficulty in saving themselves from the hot pursuit; the plan now resorted to for safety by those who find themselves in the midst of a burning prairie being to take up a position at any spot, and cut the grass for some distance around them, the fire when it makes up with them taking the circuit of the cleared spot, and thus leaving them scatheless, but uniting again after it passes them into one long zigzag belt of flame, licking up everything that is combustible in its course.

Before leaving the Mississippi valley, it may be as well to take a rapid glance at the agriculture and agricultural interest of America. In doing so I have no intention of entering into a disquisition upon practical farming; my sole object being to give the reader, from this the capital of the chief agricultural region of the country, a bird's-eye view of this all-important branch of American industry.

In the broadest sense of the term, the agricultural products of America comprise wheat, Indian corn, rice, barley, rye, oats, cotton, tobacco, potatoes, turnips, flax, hemp, sugar, indigo, fruit, and grasses of all kinds. To these may be added live stock, which are to all intents and purposes, an agricultural product. The different products here enumerated are

order in which they are named. In none of these
States is cotton the exclusive, but, in the four principal cotton-growing States, it is the staple product.
To these Florida may be added, although its annual
yield is not yet large. In the Carolinas and Georgia
rice is produced to a great extent from the low
marshy grounds of the coast, as also in the coast
districts of Florida, Alabama, Mississippi, and Louisiana. Rice has now become a leading article of
export from the South. The extent to which Indian
corn is cultivated in these States has already been
hinted at; nor is wheat altogether neglected, small
quantities of it being raised in the upland districts of
the interior in most of them. We have also already
seen how far in Virginia, wheat, and in both Virginia
and Tennessee, Indian corn and tobacco, compete
with cotton in the annual produce of these States.

The cultivation of the sugar-cane and the manufacture of sugar in the United States is chiefly, if
not exclusively, confined to the State of Louisiana.
The entire yield of this article in 1844 was computed
at upwards of 126 millions of pounds, of which upwards of ninety-seven millions were produced by
Louisiana alone. The remainder was chiefly raised
and manufactured in Georgia and Florida, there being
now every indication that sugar will yet be the great
staple product of the latter. The sugar-growers, as
a class, differ in this important particular from their
fellow agriculturists, that they join the manufacturers of the North in the cry for protection. In this
they cannot avail themselves of the flimsy pretext,
so prominently put forward by our colonial interests
and their parliamentary abettors, that one of their
objects in seeking to limit the use, if not entirely to
prohibit the introduction, of slave-grown sugar, is to

discountenance slavery and the slave-trade. Louisiana
cannot allege that one of her objects is to discountenance slavery, for her own sugar is produced by
slaves as much as is that of Cuba or Brazil. And
so long as the internal slave-trade continues in the
United States, enabling Louisiana to increase her
number of slaves by importations from the neighbouring States instead of from the coast of Africa,
she cannot, with any very high degree of consistency,
aver that her cry for protection is partly based upon
a desire to put down the slave-trade. Her object in
taking the part which she does take on the commercial question, is identical with that of those with whom
she is in league, to secure by legislative enactment a
higher profit to capital invested in a particular pursuit
than it would otherwise realise, or than capital otherwise invested would produce ; and this at the expense
of the whole body of consumers.

What an almost inexhaustible source of wealth is
there to the Republic in this variety of climate, and
this vast extent of fertile surface! With a few exceptions, such as the rocky tracts of New England, and
the light sandy plains of New Jersey, the whole area
of the country, from the Lakes to the Gulf, and from
the Atlantic to far beyond the Mississippi, is highly
productive. Even the salt marshes on the sea-shore
are capable of being turned to the most profitable
account. In many districts of an upland character,
the soil, after having been used for some time,
requires to be manured, as it does in Europe, to
renovate it. But in others, particularly in the case
of the bottom lands on the great rivers, and of valleys
well irrigated, and where the soil is rich and deep, no
manure is required. In innumerable instances has it
been worked for years in the valley of the Mississippi

CHAPTER III.

WE left St. Louis on our way up the Ohio after a sojourn of some days, during which we made several excursions to different points on both banks of the Missouri. It was at an early hour in the morning that we left; and as the day, which was exceedingly fine, wore on, nothing could look more lorldly than the Mississippi, as, after receiving the immense and turgid volume of the Missouri, it rolled swiftly on between its variegated and imposing banks—which were, on the average, about a mile apart—to take tribute from the Ohio. The bluffs on either side, with their ever-changing and fantastic forms, were to me a never-failing source of amusement and interest, particularly when I beheld them crowned by some lone hamlet or village, which the forest behind seemed to be pushing into the stream. They looked like the advanced posts of civilization in the heart of the wilderness. The main body is rapidly following up, the invasion can no longer be resisted, and the shadow of coming greatness is already forecast upon the land.

I have in a former chapter alluded to some of the peculiarities of the vanguard of the invasion. Theirs is a rough and an adventurous life, and if they are not themselves rough when they undertake it, they soon become so in prosecuting it. The pioneers in the foremost line are the most adventurous and restless of all, contracting habits in their unremitting war with nature which completely unfit them for the restraints of civilized life. The consequence is, that they fly the approach of anything like conventionality, or a settled form of society, pushing their way further and further into the forest, as permanent settlements spring up behind them. Those who follow differ at first but little from their predecessors, except in their preference for a fixed over a migratory life ; and even they are restless to a degree, as compared with the settled habits and the fixity of residence which characterise a more advanced state of society. It is chiefly this class that reduces the wilderness to cultivation, and constitutes the great agricultural body of the West. They are ready for a change of residence whenever it may appear to be a good speculation, and not, as is the case with the others, simply to enjoy, in a state of semi-barbarism, a species of romantic independence in the woods.

We had not been long afloat ere I discovered that we had several excellent specimens of the second, or settling class, on board. One particularly attracted my attention, from his enormous bulk, faultless proportions, free and easy air, and manly bearing. He was not over thirty, and was dressed in a kind of green pilot-cloth coat, although the weather was oppressively warm, his black hair falling in careless curls from under a small cap, over his face. His complexion was much lighter and clearer than that of the great majority of the Westerners, who, from the miasmas engendered by the extent to which vegetable decomposition is still going on in the soil, have generally a dark, sallow, bilious fever and aguish look about them. He had a small black eye, as quick and restless as that of a ferret. Nothing seemed to escape his observation. He first made himself familiar with every thing on board, then with everybody, and lastly gave his attention almost exclusively to external objects. Every glance which he bestowed upon you had the deep prying curiosity of a first look about it; and you could see, as his eye roved over every object from the deck to the horizon, that the mind kept up with it. He had nothing of the quiet, brooding melancholy and cunning look of the genuine Yankee about him; for whilst he observed everybody and everything, he did not seem anxious to escape observation, or to lead the judgment astray in attempting to fathom him. As he paced the deck with a confident, though by no means insolent air, I watched him for some time with the interest which attaches to a fine specimen of a noble race of animals, my admiration being divided between his herculean proportions and his manly, independent bearing. It was not long ere I got into conversation with him, although to do so I had to make the first advances. I found him shrewd, intelligent, communicative, and inquiring. He was a red-hot Oregon man, and almost gnashed his teeth with rage when he spoke of the treaty which had

been just signed by the " traitor " Polk. He had
made up his mind to reap glory in as yet unfought
fields in Canada; and being disappointed in that quarter, was now on his way to Washington, in the hope
of getting a commission, which would enable him to
vent his wrath upon the Mexicans. Having missed
all Oregon, he was now for all Mexico, and saw no
reason under the sun why a Spaniard should be left
on the northern side of Panama. The isthmus, the
north pole, and the two great oceans, were in his
opinion the only boundaries which the Republic
should recognise. He was a fiery specimen of the
fieriest Democrats, with whom the North-west
abounds—one of the " Now or Sooner " party, who
are not only carried away with the most magnificent
visions of the destinies of the Republic, but are
desirous of at once realising them.

An eclipse of the sun was that day looked for,
between one and two o'clock ; and as the hour approached I drew near to a group of negroes, who
were grinning and chattering near the bow of the
boat, each with a piece of smoked glass in his hand,
through which to observe the expected phenomenon.
On getting within reach of their voices, I found them
engaged in a biblical discussion, the controversy
hinging upon the proper meaning of the phrase, " Ho,
ye that thirst," occruring in the Prophecies. The
most loquacious amongst them, who seemed to be the
oracle of the group, held that it was chiefly applied to
those who were engaged in the cultivation of cotton
and Indian corn; the *hoe* being the principal imple-
ment used by those so occupied. Contrary as it
might seem to all experience, the exhortation addressed to those thus employed was, to *hoe* away when

they felt thirsty, that they might forget their thirst. He was indebted for this lucid interpretation to the overseer of a plantation in Alabama, on which he had been for some years a slave. It was the custom of the overseer to collect the negroes every Sunday evening, and read the Bible to them; but it appeared that, no matter from what other parts he read, he always concluded by referring to those texts which enjoined upon servants the duty of obeying their masters in all things, and showed that as a reward for working hard, the harder they worked, the less inconvenience they would feel from thirst; for such was the interpretation which he always put upon the text, "Ho, ye that thirst." This explanation was followed by a look of incredulity, which passed round the group, and drew from the speaker himself a confession that although he had often practically tested it, his experience had invariably belied the interpretation.

Shortly afterwards the eclipse, punctual to its time, commenced. It was but partial in the latitude wherein we beheld it, scarcely one-half of the sun's disc being obscured. It lasted altogether about two hours, and gave rise to many sapient and philosophic observations amongst those on board, particularly our coloured friends.

"What makes de 'clipse, Massa Gallego?" asked one of the group, addressing himself to the oracle.

"S'pose I 'splain it, Jim Snow, you no und'stand it den," replied Mr. Gallego; "but, for de sake of de oder jin'lemen I'll give you de philosophic cause of de phenomenon."

"Go it, Massa Gallego," the rest cried in chorus, exposing their huge white teeth, as they grinned almost from ear to ear.

" Well, you see," observed Mr. Gallego, encouraged by this manifestation of confidence in his attainments, " de sun is a movin' body, and so is de airth, and so, for dat matter, is de moon."

" Well," cried they all in expectation.

" So you see," continued Mr. Gallego, with all the dignity of a professor, " de sun come between de circumbular globe and de moon, and then de diameter of de moon fall upon de sun, when dey are all in de conjunctive mood."

" Well," cried his audience again.

" Well," said Mr. Gallego, in a tone of displeasure, " what are you well-ing at ? Don't you see how it is ? I can't give you no more than a 'splanation. I can't give you brains to und'stand it, no how."

" 'Cause you haven't got none to spare ; yhaw, yhaw !" said Jim Snow, bending almost double, that he might laugh the more heartily.

" Get out, nigga !" said the others, who were as little satisfied with the explanation as Mr. Snow was, but who attempted to impose upon each other by rallying round the professor, whose dignity had been grievously wounded, as was evident from the manner in which he stood, with his lips in a frightful state of protrusion, his nostrils dilated, and his eyes rolling about like those of a duck in a thunderstorm.

" Well, I no und'stand it, dat's all," said Jim Snow, deprecating the rising wrath of the company.

" Who said you did, nigga ?" said two or three of them, who on account of their superior nervous organization, had by this time been wrought up into a towering passion.

" Didn't he say dis here globe was circumbular ?" asked Jim in self-defence.

"So it is," said one of the group; "you'll not be believin' next that dey catch de pickled herrin's in de sea."

"I tell you it isn't; de globe is as flat as my hand," replied Mr. Snow.

"Neber mind him," said the professor, quelling the gathering tempest; "you might as well expect a kyow's tail to grow up'ards as that 'ere nigga to larn anything."

"If de globe is round," continued Mr. Snow, "how do de people live on de under side? Dey must stand on their heads I reckon."

"Dey live inside, you brack brockhead," replied the professor, turning round upon his heel, to put an end to the discussion.

Jim felt abashed. He was not prepared for this mode of carrying what he had evidently regarded as his strong point. His unbelief was shaken, but instead of being welcomed back into the fold, he was hissed out of the company, as a punishment for his infidelity.

When I got upon deck next morning, we were entering the Ohio. It was, at one time, intended to build a city at the confluence of the two streams, which, had it started into being, would have been a formidable rival to St. Louis. The chief obstacle in the way of the project was, that the site on which the town was to rest was very frequently under water. Cairo was to have been its name, but it by no means follows that because one Cairo can stand ankle deep in the sands of the desert, another could do so up to the knees in the marshes of the Ohio. For the present, therefore, the Cairo of the West is a mere phantasy; but that the rising exigencies of the region will, ere long, conjure into being an important commercial depôt near the mouth of the Ohio, can scarcely admit of a doubt.

The valley of the Ohio, which is merely a feature of that of the Mississippi, comprehends a large section of Illinois, the greater portions of Kentucky, Ohio and Indiana, a small part of Tennessee, and those districts of Pennsylvania and Virginia which lie west of the Allegany chain. It is irrigated by a magnificent river system, the Ohio being the main stream into which the whole valley is drained; its chief tributaries being the Wabash, which enters it on the north, and the Cumberland and Tennessee, which join it on the south bank. These, and other tributaries of the Ohio, are navigable by steamers for considerable distances, the Wabash in particular being so for about 300 miles during the greater portion of the year.

For a long distance up, the average width of the river appeared to be from three-quarters of a mile to a mile. Its current is scarcely so impetuous as that of the Mississippi, and its volume, except when it is in high flood, is as clear throughout as I observed it to be on its entrance into that river. The banks on both sides, particularly the southern bank, are undulating and picturesque, but there is a total absence of the bluffs, which form so prominent a feature in the scenery of the Mississippi. For almost the entire way up to Louisville, which is 380 miles from the junction, both banks are, with but occasional exceptions, shrouded, to the water's edge, in the dark, dense forests of the West. The prairie land of Indiana and Illinois does not extend to the Ohio. There is a flat strip of land on both sides of the river, more continuous on the Kentucky, than on the other bank, which intervenes between the river and the woody undulations which skirt it; this strip, consisting of rich deep alluvial deposit, is generally inundated when the Ohio is in very high flood.

We had nearly completed the third day after our departure from St. Louis, when, at early morning, we arrived at Louisville, the largest and handsomest town in Kentucky. It is built at the point at which occurs the chief obstacle to the navigation of the river, that which is known as the rapids of the Ohio. These rapids are trifling as compared with those which occur in the course of the St. Lawrence, extending over only two miles, and not falling much above ten feet per mile. When the river is full, the impediment which they offer is not so great as when the water is low. A short canal has been constructed around them to avoid the difficulty.

Intending to pass a day here, we immediately landed and took up our quarters in an excellent hotel. The town is well built, spacious, and pleasant, and has a thriving, bustling, and progressive look about it. The population is now about 35,000, to which it has increased from 500, which was all that it could muster at the commencement of the century.

The world has rung with the fame of Kentucky riflemen. Extraordinary feats have been attributed to them, some practicable, others of a very fabulous character. For instance, one may doubt, without being justly chargeable with too great a share of incredulity, the exploit attributed to one of their " crack shots," who, it is said, could throw up two potatoes in the air, and, waiting until he got them in a line, send a rifle ball through both of them. But waving

E 2

all question as to these extraordinary gifts, there is
no doubt but that the Kentucky riflemen are first-
rate shots.   As I was anxious to witness some proofs
of their excellence, my friend D——— inquired of the
landlord if there were then any matches going on in
town.   He directed us to a spot in the outskirts,
where we were likely to see something of the kind,
and thither we hied without loss of time.   There had
been several matches that morning, but they were
over before we arrived on the ground.   There was
one, however, still going on, of rather a singular character, and which had already been nearly of a week's
standing.   At a distance of from seventy-five to a
hundred yards from where the parties stood, were two
black cocks, pacing about in an enclosure which left
them exposed on the side towards the competitors.
At these two men were firing as fast as they could
load, and, as it appeared to me, at random, as the
cocks got off with impunity.   On my observing to
Mr. D——— that, although I was no " crack shot,"
I thought I could kill one of them at the first fire, he
smiled, and directed my attention to their tails.   One
indeed had scarcely any tail left, unless two solitary
feathers deserved the appellation.   On closer inspec-
tion, I found a white line drawn in chalk or paint on
either side of the tail of each, close to the body of the
bird, and each party taking a bird, the bet was to be
won by him who first shot the tail off his, up to the
line in question, and without inflicting the slightest
wound upon its possessor.   They were to fire as often
as they pleased, during a certain hour each day,
until the bet was decided.   One of the competitors
had been very successful, and had accomplished his
object on the third day's trial, with the exception of

the two feathers already alluded to, which, having had a wide gap created between them, seemed to baffle all his efforts to dislodge them. What the issue was I cannot say, for at the close of that day's trial it remain undecided.

Next day, we proceeded on board one of the many steamers calling at Louisville, and set off for Cincinnati, 120 miles further up the Ohio. The river differed but little in its aspect, as we ascended it, with the exception, perhaps, that the further up we proceeded, in other words, the further east, or the nearer the older States we went, the settlements on its banks were larger and more frequent, and indicated a higher stage of advancement than those below. The same difference was all along observable between the two banks, and has already been adverted to as existing between Virginia and any of the northern States. Whilst the one side presented every appearance of industry, enterprise, and activity, a sleepy languor seemed to pervade the other, which was not a mere fancy resulting from a preconceived opinion, but real and palpable. The Ohio, for almost its entire course, separates from each other the realms of freedom and slavery. It runs for a short distance within the limits of Pennsylvania, dividing for the rest of its course the States of Ohio, Indiana, and Illinois from Virginia and Kentucky. Taking into account the windings of the river, the Ohio coast of the last mentioned State is upwards of 600 miles in length.

I was somewhat disappointed by the appearance presented by Cincinnati from the river. Considering that as yet this is the capital of the West, being the largest city west of the Alleganies, I was led to expect a more imposing front than it presents to the Ohio,

on the north bank of which it is built, not far from
the south-west angle of the State of Ohio.   We landed
and stayed in the city for two days, during which we
had ample opportunity of inspecting it.   It is very
pleasantly situated on two plains of different elevations, the lower being a considerable height above the
river, and about fifty feet lower than the higher one;
both being skirted immediately behind the town by a
range of low hills, which seem to hem it in between
them and the river.   It appears, therefore, to be
cramped for room, very like Greenock, on the Clyde;
but the bowl in which Cincinnati stands will contain
a much larger population, without there being any
necessity for its invading the hills, than it is likely to
contain for many a day.   The elevated grounds are
already occupied by many residences most charmingly
situated, from most of which the town appears to
great advantage.   When seen from the hills behind,
Cincinnati amply atones for its rather shabby appearance from the river.   When in it the town is not
only passable but elegant, particularly the bulk of it
lying back from the stream.   The streets, which generally intersect each other at right angles, are very
close together, of moderate width, well shaded with
trees in some instances, and well paved in almost all.
The suburbs are somewhat scattered, and though
they appear straggling, are laid out upon a regular
plan, which can be traced by a little observation, and
which will preserve in its future increase the regularity which now characterises the city.   It is not
the capital of the State, and its public buildings are
therefore exclusively of a municipal, literary, and religious description.   None of them are large, but several,
particularly some of the churches, are exceedingly

chaste and elegant. The bulk of the better portion of the city is built of brick, with here and there some edifices of stone. The progress of Cincinnati has been most rapid, and affords one of the best exemplifications which the tourist meets with, of the celerity with which flourishing communities are conjured into existence in the New World. In the year 1800, its population did not exceed 750 souls. It is now equal to that of Aberdeen or Dundee, being about 60,000. It has thus, in less than fifty years, increased its population eighty-fold! It is one of the most orderly and industrious, and, for its size, one of the wealthiest towns in the Union; and it is much to the credit of its inhabitants that, in addition to what the State has done for education, their city abounds with evidences of a munificent liberality on their parts, with the view of still further promoting it. The stranger must indeed be fastidious who is not very favourably impressed by Cincinnati, both as regards the moral and physical aspect in which it presents itself to him.

We had been already nearly four days afloat since we left St. Louis, but were yet fully 400 miles distant, by river, from Pittsburg, our destination. The boat in which we left Cincinnati for the latter place was of smaller burden and draught than any in which we had yet been. When the summer droughts are protracted, the river, in its upper portion, sometimes becomes very low; and there are points in its channel which, on such occasions, it is difficult for even the smallest steamers to pass. There had been copious rains, however, for some days previously amongst the hills to the north-eastward, so that we anticipated no difficulty in this respect.

Amongst my fellow-passengers to Pittsburg was a Scotch emigrant, who had been settled for about five years in Ohio. He was not above thirty-five years of age, and seemed to overflow with enterprise and shrewdness. He was quite a character, and proud to a degree of the position in which he then stood, when contrasted with the obscurity of his early life. We had not been long in conversation together when he favoured me with the following bit of biography.

"I was born in Paizla," (Paisley,) said he, "where my father was a weaver body. My mither died when I was very young, and nothing would suit father but to marry again. My step-mither did na behave weel to me; she never let me eat wi' themsells, but always gave me my parritch at the door-cheek. Man, but I did na like that at a'. I was apprenticed to the weavin' mysel, but I thought I was born for better things, and partly to push my fortune, and partly to gie my step-mither the slip, I ran awa ae Friday efterneen about four o'clock; leavin' my work just as it was. I was but fourteen year then; and where div ye think I gaed."

"It would be difficult to guess, I am sure," replied I.

"Div ye ken Dunkeld?" he inquired.

"Right well," I rejoined, "one of the loveliest spots in all Scotland; charmingly situated upon the Tay, amongst the first ridges of the Grampians, as you approach them from the noble carse of Gowrie."

"Ay, I see ye ken it weel," continued he. "D'ye happen to know the Athol Arms in Dunkeld?"

"I do," replied I, "and a very excellent and comfortable house it is."

"Weel man," said he, "I was a wee bitts (boots) there for twa year. I then got tired o't and gaed awa to Glesgy (Glasgow), where I was a waiter for four year more."

"What did you do next?" I asked, getting somewhat interested in his story.

"I then," he continued, "went aboord ane of the Glesgy and Belfast steamers, where I was a steward for seven year, and after that I became travellin' agent for a speerit firm in Belfast. You see I was aye loupin' up as I thought I should, when I left the weavin'. After travellin' aboot for mair than twa year, wi' samples o' a' sorts of speerits, manufactured and sold by my employers, I packed up my things, and having saved a little money, came to this country. I came to Ohio almost as soon as I landit, and settled near Columbus, where I have a large farm, well cleared and stocked. I'm noo goin' to turn the knowledge I got in Belfast to some accoont, by setting up a whisky still—and I'm just on my way to Pittsburg for the apparatus."

"Are you married?" I asked him.

"Hoot aye man, for mair than four year back," he replied. "To get a wife was ane of the first things I did, after gettin' my farm. It's nae here as it is in Scotland, where there's mair mous to fill than there's bread to fill them wi'. The sooner a man gets married here the better, always providin' he's nae a mere striplin'. Eh man," he continued, after a moment's reflection, "if the poor Paizla weavers, that are starvin' at home, only kent what they could do here, wi' a little industry and perseverance, it's mony's the ane o' them would come awa' frae that recky, poverty-stricken hole, which would leave it a' the better for sic as were left behind."

"If instances of success like yours," observed I,

E 3

" came to their knowledge, I have no doubt but that it would stimulate many of them to follow your example. But the worst of it is that the majority of the poor with us shrink from emigration, their ignorance of what it really means investing it with vague and undefined terrors to them. There is no lack of demagogues to profit by this ignorance, and identify emigration with transportation. The poor are thus abandoned to the mercy of false teachers, instead of being taught by those who have it in their power to instruct them aright, that emigration, if the emigrant is frugal, industrious and persevering, is but a means of exchanging misery and privation at home for comfort and independence amongst one's own kindred and countrymen elsewhere."

" Your government and your rich folks have much in their power," he observed, " both in the way of instructin' the poor man how and where to emigrate, and aidin' him to leave the country, if he is so disposed. The consequences of their neglect to do so will yet recoil with terrible severity upon themselves."

Our conversation here dropped for a while, but it was long ere I could divest myself of the reflections to which its concluding portion gave rise. What wealth, what resources were around me, and at any moment within the compass of my vision, running to waste for want of a sufficient population to turn them to profitable account! What a field for the teeming multitudes of our overstocked districts! Why were they not there, enjoying ease and plenty, instead of jostling each other for a precarious subsistence at home? To what is our social system tending? Our daily national life is a daily miracle. Great as is our absolute wealth, and great as is our credit, yet as a nation are we not constantly living from hand to mouth? Derange the system by which we subsist, and the evil consequences are immediately felt. Increasing resources are relied upon as the means of ultimately relieving us from our difficulties; but as our resources increase, and as our wealth augments, our poverty also exhibits itself in more enlarged proportions. As the fabric of our national greatness towers more and more to heaven, the shadows which it casts over the landscape become deeper and more elongated. We present an imposing front to the world; but let us turn the picture, and look at the canvass. One out of every seven of us is a pauper. Every six Englishmen have, in addition to their other enormous burdens, to support a seventh between them, whose life is spent in consuming, but in adding nothing to the source of their common subsistence. And daily does the evil accumulate, and daily do we resign ourselves to it, as if it were irremediable, or would some day subside of its own accord. But the river that is always rising must, at last, overflow its banks; and a poverty which is constantly accumulating must yet strike with a mortal paralysis the system which has engendered it. There may be many cures for the evil, if we could or would hit upon them. If emigration would not prove itself a cure, it would at all events operate as a palliative until a cure could be devised. But our State doctors will not prescribe it. It would be a new-fangled treatment, and would not accord with precedent. Better spend millions a year in keeping up a nucleus for increasing poverty at home, than a few millions for a few years in wholly or partially dissipating the evil. The poor we must have always with us, and so we keep as many of them about us as we can. It is true that we have colonies, ships, and money—a redundant population at home, and vacant territories abroad—true that we have a large number here, who, for want of employment, are necessarily preying upon the industry and the energies of others, and that our colonies only want people to make them extensive markets and powerful auxiliaries to us. All this is true; but to fill the colonies and relieve the mother country, is no part of the duty of the government. It cannot interfere with private enterprise. In other words, poverty is expected to spirit itself away. The government will do nothing on an adequate scale, to invigorate the extremities, whilst it leaves a cancer to prey upon the very heart of the empire. What is to be the end of all this? It may be postponed for some time to come, if none of the sources of our national life are dried up. But let our trade receive a rude shock in any quarter, and the impending catastrophe will precipitate itself upon us in an hour.

Next day, having left Kentucky over-night behind us, we were sailing between Ohio and Western Virginia. The country on either side was now more broken and hilly than any portion of it lower down the river, and gave token, every step that we advanced, of our nearer and nearer approach to one of the great mountain systems of the continent. But, as yet, the undulating surface in no part rose to the dignity of a mountain, being composed of a succession of small hills, which appeared capable of cultivation to the very top. At the close of the second day, however, as we approached the frontier of Pennsylvania, the land began to heave itself up in larger and more abrupt masses from the plain, whilst here and there could be faintly traced along the eastern horizon the distant crests of the Alleganies. Thus seen at a great distance, they looked like purple clouds afloat in a sky of azure; and delicious to me—after being for some weeks accustomed to nothing save the level and monotonous lines of woodland and prairie, which constitute the chief features in the scenery of the great valley—were these first and far-off glimpses from the west of this glorious mountain-chain.

Owing to some detentions by the way, it was the afternoon of next day ere we reached Pittsburg, when,—after a journey of 1,100 miles from St. Louis, and no less than 2,300 from New Orleans, and all on the bosom of two great rivers, passing through an enormous region unsurpassed in fertility and unequalled in its natural advantages, and flowing through almost interminable tracts of forest and prairie, and by flourishing cities, rising towns, and sweet smiling villages,—I stepped ashore on the right bank of the Monongahela.

Pittsburg, the capital of Western Pennsylvania and the chief seat of western manufacture, is, commercially speaking, most advantageously situated on the peninsula formed by the confluence of the Allegany and Monongahela rivers, which here unite to form the Ohio. It is thus in direct communication with the whole valley of the Mississippi, and with the Delaware and the Atlantic, by means of the Pennsylvania canals. It will also soon have a continuous water communication with the Great Lakes, by means of the Genesee valley canal, already partly constructed and designed to unite the Allegany River with the Erie canal at Rochester in New York. The chief port of Pittsburg is on its Monongahela side, where, throughout the year, is the greatest depth of water. It is connected with the opposite shores of both rivers by means of stupendous bridges, leading to the different suburbs by which it is surrounded, Allegany city, the principal one, being on the right bank of the river of that name. The town, partly owing to its position, is very compactly built; and some of its public buildings, which are substantial and elegant, are well situated for effect upon the rising ground immediately behind it. The country around is broken and hilly, the hills containing inexhaustible stores of the bituminous coal, which Pittsburg uses to such an extent in connexion with its manufactures. It is termed by its inhabitants the " Sheffield of the West," from the similarity of its manufactures to those of that town. In one thing it certainly resembles Sheffield—in the dingy and sickly character of the vegetation in its immediate vicinity; the fresh green leaf and the delicate flower being begrimed, ere they have fully unfolded themselves, by the smoke and soot with which the whole atmosphere is impregnated. Both iron and coal are found in vast abundance in its neighbourhood, from which the character of its industry may be inferred. It has furnaces for the manufacture ;of cast-iron; it has bloomeries, forges, and rolling-mills; and carries on an extensive manufacture of cutlery, hardware, and glass. The aggregate amount invested in manufacture in Pittsburg comes close upon three millions sterling. In 1800 its population was considerably under 2,000, it is now 30,000. Its future growth is sufficiently typified by its past progress. Its canal communication with Philadelphia is interrupted by the Alleganies; but the broken link is supplied by a short railway, which crosses the mountains by means of stupendous inclined planes and heavy tunneling. The canal-boats are generally divisible into three parts, each part being capable of floating by itself. When they reach the mountains they are taken to pieces, placed upon trucks, and carried across by railway, when their different parts, being once more launched and afloat, are hooked together, and thus again forming one boat proceed on their journey.

Pittsburg being situated on the confines of one of the greatest mining districts in the United States, no better opportunity can offer itself of taking a very general and rapid glance at the mineral resources, and the mining interests of the Union.

There is no country in the world possessing a greater abundance, or a greater variety, of mineral resources than the United States. There is scarcely a known mineral existing that is not found somewhere, and in greater or less quantity, within the limits of the Republic. We have already seen the extent to which the gold region stretches from North-east to South-west, although it may not have been found very productive at any particular point. But if any credit is to be attached to the accounts which now reach us from California, the Union has, by its recent acquisitions from Mexico, added to its territories an auriferous region, as rich as any yet discovered in the world. The silver mines of the continent seem to be chiefly confined to the countries lying to the west of the Gulf of Mexico, although this metal is found in small quantities in some of the Southern States. Quicksilver, again, is found in great abundance, and in different combinations, in the northern and western districts, that is to say, in the neighbourhood of the lakes. Although copper is found elsewhere, it is only in the neighbourhood of Lake Superior that it has as yet been discovered in any very large quantity. During the mania for copper mining, which a short time ago pervaded both the United States and Canada, some parties either purchased or leased enormous tracts, in some cases consisting of several miles square, for the purpose of carrying on operations from which they expected immediately to amass colossal fortunes. But like most of those in too great haste to be rich, their splendid visions have to a great extent faded. There is no doubt, however, but that there is an abundance of copper in this region, which, when better communications are opened with so remote a quarter, will be turned to profitable account.

The continent is abundantly supplied with iron. Within the Union it is found in greatest quantity in the States of Pennsylvania, New Jersey, Maryland, and New York. There are also extensive iron mines as far south as Virginia, which are as yet but very partially worked. It is in the north-west parts of Illinois that lead is found in the greatest abundance. The ore found here is as rich as any lead ore in the world, particularly that produced near Galena, which is the chief seat of mining operations in connexion with this metal. The supply appears to be inexhaustible, and lies so near the surface that even the Indians used to produce great quantities of lead here, before the attention of the whites was drawn to the mineral wealth of the district.

Almost all the salt made in the United States is the produce of salt springs. The greatest hitherto discovered are in Onondaga county, New York. They are State property, and yield a large revenue annually to the State exchequer.

But with all this vast and varied supply of minerals, the United States would still be at a loss if they were wanting in coal, the great agent employed in bringing most of them, if not all, into practicable shape. But if there is one mineral production with which they are more liberally supplied than another, it is this. One enormous coal region, with many interruptions it is true, stretches from the southern counties of New York to the northern counties of Alabama. Coal is also found in New England and New Jersey, and vast fields of bituminous coal lie close to the surface, in the neighbourhood of Richmond, Virginia. The chief of these, the Chesterfield coal field, is worked by an English company. The whole coal area of the United States is estimated at upwards of 70,000 square miles, about twelve times the extent of the aggregate coal area of all Europe, and about thirty-five times the extent of that of Great Britain and Ireland. The coal area of the United States is nearly as great as the entire area of Great Britain. The Americans too have this advantage in working their mines, that the mineral lies near the surface, or is generally otherwise found in accessible positions.

But unquestionably the chief interest that attaches to mining operations in the United States centres in Pennsylvania. As New England is the chief seat of manufactures, so is that State the chief focus of mining industry, as it is the chief seat of mineral wealth. Fully one half, if not more, of all the iron manufactured in the United States is the produce of the mines and industry of Pennsylvania. Nor is its mineral wealth very partial in its distribution. In five out of every eight counties in Pennsylvania, and the total number is fifty-four, both iron and coal are found in abundance. The coal area of the State particularly is enormous, extending over 10,000 square miles, being about five times the extent of that of Great Britain and Ireland. It is as yet but partially worked, but what a source of wealth and greatness is here ! The coal mines of Pennsylvania are as rich as any of those in England, and the strata in most cases lie so close together, that several can be worked at little more than the cost of working one.

Not only are the Pennsylvanian mines as rich as, but they also produce a greater variety of coal than the English mines. The produce of the former is primarily divisible into two great classes, the bituminous and the anthracite coal. The great seat of the latter species is between the Blue Ridge and the Susquehanna, east of the Allegany mountains, whilst the former is principally, if not exclusively, found immediately westward of the chain. The Alleganies thus divide the two great coal fields of Pennsylvania from each other, forming the western boundary of the anthracite and the eastern boundary of the bituminous field.

The value of the bituminous coal was, of course, appreciated as soon as it was discovered ; but it was some time ere it was known that anthracite coal could be turned to the same purposes as its rival. It is now not only extensively used for domestic purposes, but also in the operations connected with smelting, and forging, and casting. Its availability in this respect materially enhances the mineral wealth of Pennsylvania.

Some estimate may be formed of the extent to which these resources will yet be applied, by glancing at that to which they have already been turned to account. For the figures which follow I am chiefly indebted to some articles which appeared in 1847 in the *Philadelphia Commercial List.* The principal development of the coal resources of Pennsylvania has been in connexion with its great anthracite coal field, that being most accessible to the markets in which coal is now most in demand. It was only in 1820 that it first appeared as a marketable commodity; and in that year the quantity sent to market on tide-water did not exceed 365 tons. For the nine years that succeeded, the average annual receipts of anthracite coal at tide-water were 25,648 tons. For the next nine years the annual average was 454,534 tons; and for the succeeding nine, terminating in 1847, it was no less than 1,283,229 tons. This rapid rate of increase in its consumption demonstrates the availability of the article, the facility with which the mines can be worked, and the growing demand for their produce.

It was long after anthracite coal came to be very generally used for domestic purposes, that it was applied to smelting and other kindred operations. Indeed, so late as 1840, there were no furnaces in Pennsylvania consuming this species of coal. There are now from forty to fifty in full operation using it, and some of these are of the largest class. Numerous rolling mills have also been erected during the last few years, so constructed as to consume it; and it is difficult to keep pace with the rapid increase in the demand for it. Since it has been brought into general use, it has more than trebled the coasting trade of Philadelphia, and, as noticed in a former chapter, the trade to which it has given rise has called into sudden existence the suburb and port of Richmond, immediately above the city, which is now the chief seat of its export. The abundance in which it is found, and the ease with which it is already worked, are evident from its cost at the mouth of the mine, which is, on the average, but thirty-five cents, or 1s. 9d. sterling, per ton.

The localities in which it is chiefly found, are what are known as the Lehigh and Schuylkill regions. The value of the coal trade to Pennsylvania, and the prospects which it appears to hold out, may be inferred from the enormous amount of money already invested in internal improvements, constructed chiefly, if not wholly, with a view to facilitating its transit to market. The Lehigh improvements, in the shape of canals, railways, &c., have cost 7,045,000 dollars, or 1,384,325l. sterling. The aggregate sum invested in improvements connected with the Schuylkill coal region, is 19,365,000 dollars, or 4,034,375l. sterling. These sums, with the cost of other improvements, not exclusively connected with the coal trade, but affording it every facility in reaching the Hudson and New York, make a total thus invested of no less than 34,970,000 dollars, or 7,285,416l. sterling.

This glance simply embraces the anthracite coal trade east of the mountains. The great bituminous region to the west, extending to the vicinity of Pittsburg, is being also rapidly developed, the enormous trade which will yet spring from it being destined to embrace the regions bordering the lakes and the valley of the Mississippi. Large quantities of bituminous coal are and will be consumed upon the sea-board; but, except where Nova Scotia and English

CHAPTER IV.

FROM PITTSBURG TO NIAGARA.

FROM Pittsburg I had the choice of several routes to the lakes, but on account of the beauty and variety of its scenery, I selected that by the Genesee valley through Western New York. My friend D——— had left me at a point on the Ohio, some distance below Pittsburg, whence he proceeded to Cumberland, where he would get upon the Baltimore and Ohio railway, which would convey him to his home. I was therefore left to find my way unaccompanied towards Lake Ontario, and proceeded after a sojourn of two days at Pittsburg, northward to Olean Point, on the border of New York, at which point the Genesee valley canal, starting from the Erie canal at Rochester, is to communicate with the Allegany River, and consequently with the valley of the

Mississippi. The portion of Pennsylvania which I had to traverse to reach this point offered to my delighted eyes the most charming variety of scenery that I had as yet come in contact with. The chief ridges of the lordly Alleganies were at a considerable distance to the east, but it is long ere the land, extending on all sides from their bases, loses its billowy aspect and sinks into the level plain. Almost the entire course of the Allegany River is through a broken and romantic country, rich both in superficial and internal resources. The hills enclose an abundance of mineral wealth in their bosoms, whilst the valleys which they bound are fertile, and in many cases beautifully cultivated. The forest in this western region of the State has as yet been but partially invaded, but every year now witnesses the rapid exposure of new areas to the sun. In many of the valleys there is the richest growth of timber of almost every variety, whilst the swelling sides of the hills are frequently enveloped in one deep dark mantle of pine. Even in America, where there is so great a glut of timber, that which borders the Allegany is valuable. Its proximity to the river renders it accessible to different markets, and taking nothing else into account, the increasing value of the timber alone is rapidly enhancing the value of the soil which it encumbers.

Proceeding from Olean to Angelica, which is but a short distance, I passed over some high ground, which would have attracted but little of my attention, were it not for the important part which it plays in the geography of the continent. Narrow though the ridge be, and unimposing as it is in point of altitude, it is here the dividing line between two of the greatest river systems in the world, separating in fact the basin of the St. Lawrence from the valley of the Mississippi. The waters of the Allegany, and other streams which rise on one side of it, flow towards the Gulf of Mexico; whilst those of the Genesee and its tributaries find their way, through Lake Ontario, to the Gulf of St. Lawrence. Descending from this important, though unobtrusive elevation, and proceeding in a north-easterly direction, I soon found myself in the charming village of Angelica, the capital of Allegany county in New York. It is close to the Genesee, and hemmed in on all sides by bold rising grounds, most of them wooded to their summits; whilst the line of its horizon is broken and undulating to a degree.

A ride of a few hours brought me from Angelica to Portage. The country between them was of the same uneven character as that which lay south of the former place. The village of Portage, although insignificant in point of population, is romantically situated on the left bank of the Genesee, just as the river enters the stupendous gorge by which it forces its way through a hilly ridge, about thirteen miles in width. Immediately above the bridge which crosses it at Portage, the Genesee is calm and tranquil as a mill-pond, but a few yards below it is broken into rapids, and goes brawling and foaming over a rocky channel, until it is lost to the sight amid the dark grey cliffs which overhang it.

The student of American geography will frequently, in tracing the streams, find the word "Portage" upon the map. It is of French origin, and denotes that, at the point where it is found, the navigation of the stream is interrupted by some impediment, which compelled the early *voyageurs* to carry their canoes

round the obstruction, until they gained a point
where the channel was again practicable.   Here was
a portage of no less than thirteen miles in length, the
navigation of the Genesee being for that distance impossible, from impediments which I now proceed to
describe.

Under the guidance of one of the villagers, I
ascended, by the main road, the long hill which rose
from the opposite bank of the river.   Having gained
the summit, we diverged to the left, into a dense
forest of pine, through the twilight formed by the
dark shadows of which we forced our way, until we
approached a thicket of underwood, through which it
was scarcely possible to pass, and which veiled every
object beyond from our view.   By this time, the
sound as of " many waters " fell distinctly upon my
ear, seeming to proceed from the right and from the
left, and from far beneath my feet.   Caution was
enjoined upon me as we pressed through the thicket,
and not without reason, for we had not proceeded
many yards ere I could perceive, through its tangled
trellis-work of boughs, that a chasm intervened between
us and a cliff opposite, which was within two hundred
yards of us.   We were on a level with its weather-
beaten brow, of which we got but an occasional
glimpse, as the wind swayed the dense foliage to and
fro.   As we cautiously advanced, the naked and perpendicular wall of rock opposite seemed to descend to
an interminable depth.   We were soon upon the
verge of it next to us, but there still appeared to be
no limit to the depth of the chasm.   The thick underwood bent over the precipice, so as to conceal the
greater portion of what was beneath from our view;
and it was only by climbing a half-grown pine that

we could fairly overlook it. A scene of indescribable grandeur then burst upon my sight. The chasm for nearly three-quarters of a mile in length lay unveiled at my feet. It was only here and there that I could get a sight of the river, which was bounding from rock to rock, and covered with foam. It was more than 400 feet beneath me, and although its course was in reality exceedingly rapid, yet seen from such a height, it seemed to crawl along like a wounded snake. It was lined on either side, and its channel interrupted by masses of loose stone, which had fallen one after another from the huge cliffs which rose in gloomy grandeur over its bed, casting their ponderous shadows upon its agitated surface. The cliffs were as perpendicular as a wall, and the horizontal strata of sandstone, of which they were composed, had about them the regularity and the appearance of mason-work. The rich foliage swept, like soft hair, in waving masses over their beetling brows; its warm shades of green forming a pleasing contrast with their cold grey sides. They stood so close to each other that two persons standing on either side of the cleft could converse together with but little extra effort of the lungs.

On listening more attentively I discovered that the sound which proceeded from the rapids below was accompanied by a hoarser and a deeper note, which seemed to issue from behind a slight bend in the gorge to my left. On inquiring into the source of this, my guide, informed me that it arose from the falls, which were visible from a point about a quarter of a mile above us. Emerging from the thicket, we were not long in reaching it; and on approaching its verge, two magnificent cataracts broke at once

upon my startled vision.   The upper fall was nearly
a third of a mile from where we stood, and about
half a mile below the point at which the river
entered the gorge at Portage.   I could see but little
of the stream above it as it swept suddenly round to
the left; but the portion of it visible was broken into
rapids and white with foam. This fall is about seventy
feet in height.   Immediately below it the river is
deep and tranquil, continuing so until it comes within
a few yards of the second plunge, which is preceded
by a short rapid.   The second is the more stupendous
fall of the two, being 110 feet in height, and overhung on either side with frowning masses of rock.
Directly above it, the bank on which we stood lost
its precipitous character, being covered with timber,
and shelving rapidly down to the edge of the river.
We descended, and found a ferry between the two
cataracts.   Hiring the ferry-boat, we were rowed to
the upper fall, which, when closely approached, resembled the three sides of a rhomboid, with the
longest sides in the direction of the stream.   We
sailed cautiously within its fearful walls, and, when
tossed about by the boiling caldron at its feet, were
completely surrounded on all sides but one by the
falling waters.   Looking out, as it were, from the
embrace of one cataract, we could trace, through
the narrow gate by which we had entered, the
placid course of the river until it reached the line at
which it took its plunge to form another, when we
suddenly lost sight of it.   Having dropped down to
the ferry, we then crossed the river to a point where
there was a short break in the other bank, near which
were a saw-mill and several wooden huts. After
scrambling up the bank we came to a high road, some

distance back from the river, which we pursued for about two miles, taking the course of the stream which was on our right. We then crossed some fields, and once more approached the chasm.

The bank here was not perpendicular, but it was exceedingly steep and densely wooded—the topmost branches of one tree waving around the roots of another. Looking down, nothing was visible save a mass of foliage; but I was anxious to descend, for the roar of another cataract was already in my ear. But it was no easy matter to do so, from the steepness and loose slimy character of portions of the bank. By the aid of roots and branches, to which we clung, we managed to descend for nearly two hundred feet, when we suddenly emerged upon the bed of the river, which was one mass of rock. We stood upon a broad platform, formed by a lofty ledge, which lay across the course of the stream. The water, however, had worn for itself a narrow channel on this ledge, close to the opposite bank, which was quite bare and precipitous for some height, after which it slanted off and was covered with wood like that which we had descended. Pouring through this channel, as through a funnel, the raging current was dashed against a rock which projected at a right angle from the bank, and which turned it suddenly to the left, to fall over another ledge about ninety feet high, which lay not across the river, but parallel to the two banks. When in full flood the stream dashes furiously over the ledge on which we stood, taking a perpendicular plunge into the abyss below of nearly two hundred feet. For the rest of its way through the gorge, the agitated Genesee is a succession of rapids, overhung alternately with steep wooded banks and stupendous precipices. About half a mile below the third and last fall, the cliffs rise perpendicularly on either side to a height exceeding 500 feet.

Such are the falls of Portage on the Genesee, which scarcely one traveller out of a hundred who make the tour of the Union either sees or hears of. Yet they are within little more than half-a-day's easy ride of Rochester. In magnitude they cannot of course be compared with Niagara, but in the stupendous character of their adjuncts they far exceed it.

I slept soundly after my day's fatiguing ramble, and next morning proceeded towards Rochester. The ride over the ridge was highly interesting. On my right lay Nunda valley, speckled with clearances, and on my left the gorge of the Genesee, which I could trace by the grey crags which every now and then peered over the intervening tree-tops. The road, which is exceedingly rough at some seasons of the year, was smooth and pleasant, some showers overnight having laid the dust, and the gig in which I was seated passing as softly over it as if it had been rolling upon velvet. The air was bright and clear, and on my gaining the summit of the ridge, Lake Ontario was visible far to the northward, like a deep blue line underlying the horizon. I involuntarily rose to my feet on catching the first glimpse of one of the links of that great freshwater chain, which forms the most prominent feature of all in the physical phenomena of America.

After a ride of nearly two hours' duration, I approached the village of Mount Morris. For the last two miles the descent was rapid. I was now fairly in the valley of the Genesee, which extended to the right and left as far as the eye could reach. The

Genesee enters the valley at right angles, a little below Mount Morris, emerging from between two majestic cliffs, similar in character and grandeur to those which rise over it at Portage. A huge dam has been constructed here in connexion with the Genesee Valley Canal, which crosses the river at this point, and passing by Mount Morris, proceeds by Nunda valley to the falls, past which it is carried by excavations and tunnels along the very verge of the precipice to Portage, where it again crosses the Genesee by an aqueduct. The upper portion of the valley, that which lies south of the point at which the Genesee enters it, is watered by a small stream which joins it as a tributary. After flowing over the dam, the Genesee brawls along a broad stony channel until it finds the lowest level of the valley, when turning to the northward it pursues a sluggish and serpentine course through a rich alluvial deposit to Rochester.

Mount Morris occupies a beautiful position, about a third of the way up the west bank of the valley. The prospect which it commands embraces nearly the whole of the rich and fertile county of Livingstone. Although at the commencement of the century scarcely a tree of the forest had been felled in it, the greater portion of the valley between Mount Morris and Rochester is now cleared ; its two banks, which recede from the river in successive terraces, being covered with waving corn-fields, and speckled with charming and flourishing villages. The lower portions of the valley, where the deposit of rich mould is deep, are fertile to excess, being famed for their exuberance throughout the country as the Genesee flats. This favoured region is the granary of New York, and no flour is in greater repute than that which bears the Genesee brand.

Descending from Mount Morris, the road led directly across the valley. Whilst traversing the bottom lands it was for two miles as flat as a bowling-green, the wheels sinking deep into the free, black, rich mould over which I was driven. On gaining the opposite side, the road rose for some distance up the east bank of the valley, after which it turned sharp to the left, and proceeded along an elevated terrace, northward, towards Lake Ontario. It was here that the best views of it were to be had, and nothing could surpass the beauty and richness of the extensive landscape which it presented ; corn-fields and meadows alternating in rich succession along the bottom lands, and on either margin of the sluggish, snake-like stream, which lingered amongst them, whilst far up the western bank, and along that on which I was riding, the golden corn was either already cut, or waiting for the sickle. I had seen nothing in America which in appearance so nearly approximated a fertile rural district of England.

In the course of an hour I drove up to a comfortable hotel in the charming and beautifully situated village of Geneseo. After dining, I again took the road for Avon, celebrated for its mineral springs, and lying a few miles to the northward. There I again diverged to the left, and re-crossed the valley, passing the Genesee by means of a covered wooden bridge which spanned it, and pursuing my way on its left bank soon reached the village of Scottsville. Thence a ride of twelve miles, all through the richest country, and the last eight of which led by the margin of the river, brought me to the city of Rochester. It was nearly sunset, when, on gaining the top of a low hill, about a mile to the south of it, and over which the road led through a thick wood, the town burst in an instant upon my view; and few scenes could surpass in beauty that which then lay before me—the city lying below in the midst of a spacious plain, with its spires, towers, and cupolas gleaming brightly in the golden lustre of an autumn evening.

There is no other town in America, the history of which better illustrates the rapid progress of material and moral civilization in the United States, than that of the city of Rochester. In 1812, but a single log hut occupied the site of the present city. In the short space of thirty-six years it has spread over both banks of the Genesee, until it now contains upwards of 30,000 souls. Ten years hence, computing it at the ratio in which it now progresses, its population will exceed 50,000. It is now pretty equally divided between the two banks of the river, although for many years the bulk of it was confined to the west bank, which was for some time wet and marshy, but is now drained and rendered perfectly healthy. The city takes its name from that of its founder, Colonel Rochester, the numerous members of whose family have ever taken the most prominent position in the pleasant and highly-cultivated social circle which exists in it.

That which attracted the first settlers to the site of the future city, was the inexhaustible and easily available water power which the Genesee there afforded them. From the point at which it escapes from the gorge at Mount Morris, the course of the river continues sluggish and smooth until it is fairly within the precincts of the city, when it becomes once more disturbed by rapids, which are but the precursors of a still greater change.   Before reaching Lake Ontario, which is but seven miles distant, the Genesee is destined to take three additional plunges, like those which it takes at Portage, over three successive ledges of rock.   The three falls which here occur are all within the municipal limits of Rochester.   At the city the bed of the river is from two to three hundred feet above the level of Lake Ontario.   The surface of the country falls but little on approaching the lake, but the channel of the river rapidly declines, and gains the level of the lake at a point about two miles and a half below the densely-built portion of the town.   The first obvious declination of the channel occurs about a quarter of a mile above the upper fall.   The smooth current is broken by some shallow ledges of rock, and ere it has proceeded three hundred yards, becomes a foaming rapid.   In the midst of this, and upon the solid rock forming the bed of the river, stands a magnificent stone aqueduct, by means of which the Erie canal is carried across the river.   The agitated and chafing waters pour with impetuous velocity through its seven noble arches, and it forms altogether one of the finest specimens of bridge architecture in the world.   It is built of granite, and was completed about five years ago, when it replaced another aqueduct of smaller dimensions, which had been constructed of a species of red sandstone, which rapidly decomposed on exposure to the elements. Above the aqueduct is a wooden bridge, by means of which the southern portions of the city communicate with each other.   Immediately below it is another bridge, in the line of the main street of the town. From the upper bridge to the fall the rapids continue with but little intermission.   At its first great leap the Genesee here takes a perpendicular plunge of ninety-six feet, the width of the fall being about a furlong. This is decidedly the finest fall in the whole course of the river, although its adjuncts, in point of scenery, fall infinitely short of those of the Portage falls. Above it, where the city is chiefly built, the banks of the river are low, but immediately below they become lofty, rugged, and picturesque.

The extensive water power, of which the city has so largely availed itself, is furnished by the rapids and the upper fall. Almost from where the former commence, to a point a considerable distance below the latter, both banks are lined with flour-mills, tanneries, saw-mills, and manufactories of various kinds. Rochester has thus no quays upon the river, a great defect so far as its appearance is concerned. Like London, it turns its back, as it were, upon the noblest feature in its site.

Ever since its foundation the chief manufacture of Rochester has been that of flour. It is not only the principal place for the manufacture of this commodity in the United States, but also, perhaps, in the whole world. There are several mills in it which can turn out 500 barrels of flour per day, and the aggregate quantity manufactured in it last year very nearly amounted to a million of barrels. The wheat which it grinds is chiefly the produce of the fertile valley which lies behind it. Recently, however, factories of different kinds have sprung up within it, and coarse calicos, broad-cloths, and edge-tools now figure largely amongst the products of its industry. For all this it is indebted to its inexhaustible water power.

The great western line of railway, uniting the sea-coast at Boston with Lake Erie at Buffalo, is carried over the Genesee on a somewhat ricketty-looking wooden bridge, not much more than thirty yards above the fall.   Many a timid traveller shrinks in crossing it, when he looks from the gleaming rapids which are shooting the bridge with fearful velocity beneath him, to the verge of the cataract upon which he could almost leap from the train.

Between the upper and the middle fall, to which a romantic walk leads the tourist, along the precipices on either side, the river is almost one continued series of gentle rapids.   About a couple of miles intervene between the two cataracts, and the water power afforded by the rapids is available at most points.   In many places the banks are naked and precipitous, and of the same character as those of Portage, though by no means on the same gigantic scale.   At other points they slope gently down to the river, covered with grass, the timber having been cleared away from them, whilst here and there a piece of flat ground intervenes between the stream and the bank, which recedes for a short distance in an amphitheatric sweep from the water.   These spots will yet be occupied by streets, mills, and factories.   The middle fall is inferior to the other two, the plunge not exceeding thirty feet. Paper and other manufacturing establishments line the west bank immediately below it, which is one of the pieces of flat ground alluded to above.   From this to the lower fall the distance is about a quarter of a mile, the river rapidly descending between them by a series of brawling rapids.   The height of the lower fall is upwards of seventy feet, and although inferior both in height and width to the upper one, it is by far the grandest and most striking of the three.   As

designated Mount Hope, and from its highest peak, from which the timber has been cleared away, sweet glimpses of the town are caught between the tree-tops immediately below. You can almost distinguish the hum of the busy city of the living from the midst of the silent city of the dead; whilst you have within the range of your vision an impressive epitome of human life in the factory, the spire, and the tombstone.

The principal charm of Rochester is in its social circle, which is intellectual, highly cultivated, hospitable, frank, and warm-hearted. Some time previously, whilst sojourning for a considerable period in the city, I had every opportunity extended to me of mingling freely with its society; nor can the busy scenes or the excitements of life ever suffice to erase from my mind the remembrance of the many pleasant days which I have spent, or the recollection of the many friends whom I have left behind, in Rochester.

For Niagara at last! With what highly wrought anticipations did I prepare for the journey! I had a choice of routes, by railway to Lewiston and thence to the Falls, or by steamer from the Genesee to Lewiston. Anxious to find myself afloat upon one of the great lakes, I preferred the latter, and proceeded at an early hour on a fine summer morning to the upper port of Rochester, which is about half a mile below the lower fall, and nearly four miles from the lake. Descending a long and steep hill, cut with great labour and at a heavy cost along the abrupt sides of the lofty wooded bank, I reached the river, and put my luggage on board the steamer which was moored to a low wooden wharf. As I was about an hour before the time of starting, I hired a boat and dropped down to the mouth of the river, where, on its left bank, stands the village of Carthage, the lower port of the Genesee. I have seldom enjoyed a more delightful sail. The high banks which rise on either side were buried in foliage, except where, here and there, the red sandstone protruded through the rich soft moss. The channel being winding, my eye was charmed with a constant succession of pictures, until at length, on turning a low naked point on the right, the boundless volume of Lake Ontario lay rolling before me.

I landed at Carthage and awaited the steamer, which always touches at it on her way. If the original Carthage played an important figure in the wars of Rome, its modern namesake is not wholly unconnected with the military annals of America. During the last war an expedition, under the command of Sir James Yeo, landed here, and proceeded up the west bank of the Genesee, with a view to capture Rochester, which was then but in the germ. The citizens, with one exception, turned out manfully for the defence of the place, and hastily constructed a breastwork on the southern bank of a ravine, about three miles to the north of the city, and which the invaders would have to pass to attain the object of the expedition. The exception was that of an old deacon, who was as brave as a lion, but who believed that he could best serve his country's cause by remaining behind and praying for the rest, who had gone forth to fight. Whether from want of spirit on the part of the invaders, the valour of the citizens, or the deacon's prayers, has not yet been ascertained, but it is an historical fact that the expedition never passed the ravine. Sir James immediately afterwards embarked his forces again at Carthage; and if in his next despatch he was not able to say, *Delenda est Carthago,* it was because at the time there was little or nothing in it to destroy. The modern Marius sat not amongst the ruins of a past, but amongst the germs of a future town.

After a stay of five minutes at Carthage, the steamer resumed her journey, gliding into the lake from between two long parallel jetties, which form the entrance into the harbour. The sun shone brightly, not a cloud being visible above the horizon, whilst the fresh breeze which came with cooling influence from the north-west, agitated the surface of the deep blue lake. There was nothing to indicate that I had not been suddenly launched upon the wide ocean. On our left, as we steamed up the lake, we had the low shore of New York; but on our right, and behind and before us, no sign of land was visible. I tasted the water, which was pure, sweet and fresh, ere I could divest myself of the belief that it was the sea after all. I had already had ample experience of the gigantic scale on which nature has fashioned the other great features of the continent. I had traversed the plain, whose boundaries seemed to fly from my approach, and had traced for thousands of miles, the river and the mountain chain; in addition to which my mind was fully impressed with the immense size of the North American lakes; but I was not prepared for half the surprise which I felt, on actually finding, when thus afloat upon one of them, the horizon rest upon a boundless waste of waters. Violence was at once done to all my preconceived notions of a lake, one of which was that it should, at least, have visible boundaries. But the mind expands or contracts with the occasion, and so accustomed did I soon become to objects whose magnitude at first overwhelmed me, that I frequently afterwards found myself, for a day at a time, entirely out of sight of land on these fresh-water seas, without deeming the circumstance in the least degree extraordinary. Lake Ontario is the smallest of the great chain; but it extends, never-theless, for upwards of 200 miles from east to west, whilst its average width is about sixty miles. Opposite the mouth of the Genesee, it is fully seventy miles wide. Yachts and pleasure boats deck the surface of our English lakes; hostile fleets have come in collision on those of America. The waters of the latter are ploughed by the steamboat, the brig, and the schooner, in time of peace, and by the thundering frigate in time of war. In the fall of the year, the American lakes are frequently visited by disastrous tempests, when a sea runs in them which would do no discredit to the Atlantic in one of its wildest moods, and great loss of life and property is sometimes occasioned. In the early days of the province of Upper Canada, and before the introduction of steamers, the passage of the lake was made by means of schooners or other sailing craft. On one occasion a schooner-load of judges, clerks of assize, attornies, and barristers-at-law, left Toronto for Cobourg, seventy miles distant, to attend circuit. Neither the vessel nor crew was ever heard of. They had all perished in a tempest. There were not wanting those who were impious enough to deem the visitation a good riddance. To supply the void thus made, lawyers were afterwards created by act of parliament.

It was towards evening when we made the mouth of the Niagara River, which discharges the surplus waters of Lake Erie into Lake Ontario, entering the latter on its south bank, and about fifty miles below its western extremity. It is the dividing line between the different jurisdictions of Canada and New York, where the two systems stand confronting each other, which are now battling for supremacy throughout the world. There can be but little question as to which of them is ultimately to prevail, whether for good or for evil, in the New World. Neither bank is high at the mouth of the river, but both are abrupt. A fort occupies the point on either side. Over that on the left, as you enter, floats the gorgeous flag of the Union; over the other, the ubiquitous emblem of England. They are now streaming quietly in the breeze, but the times have been when they were wreathed in smoke and dragged in blood. There was no portion of the frontier which, during the last war, witnessed so many desperate and internecine conflicts, as the grand and majestic link in the long boundary which stretches from the one lake to the other.

We touched at the town of Niagara on the Canada side, lying some distance back from the river, on a gentle acclivity. Directly opposite, and on the northern shore of the lake, lay Toronto, at a distance of about thirty-six miles, its width rapidly diminishing as the lake approaches its western extremity. At the mouth of the Niagara we were but fourteen miles from the Falls, and my impatience to proceed was almost beyond control. After a few minutes' stay at the wharf, we proceeded up the broad deep river. The bank on either side became loftier as we ascended, being, for the most part, covered with timber. The current ran swiftly, but was not broken into rapids, its blistered looking surface indicating at once its depths and its impetuosity. The shades of evening were darkening the landscape as we arrived at Queenston, seven miles up, and at the head of the navigation of the river from Lake Ontario. The American town of Lewiston lay on the opposite bank of the river, but I stepped ashore, ere the steamer crossed to it, and found myself, after an absence of many months, once more on British soil.

It is easy, from either Queenston or Lewiston, to discern the *rationale* of the Falls. Both these places lie at the foot of a steep ridge, which extends, like a chain of hills, from either bank of the river, across the country. On gaining the summit of this ridge, you do not descend again into a valley on its opposite side, but find yourself on an elevated plateau which constitutes the level of Lake Erie. The Falls are thus occasioned by the surplus waters of Lake Erie descending to the lower level of Lake Ontario. The whole descent is not made by the Falls, there being a series of rapids both above and below them, those below extending for seven miles to Queenston. There the river, emerging from the ridge, as from a colossal gateway, pours with impetuous velocity into the broader and smoother channel, by which it glides into Lake Ontario. It is evident that the Falls must at first have been at the point where the country suddenly sinks to the level of that lake, in other words, at Queenston, from which, during the lapse of ages, they have gradually worn their way back to their present position, seven miles from that town. The channel which they have thus carved through the upper level is narrow, and overhung by frowning and precipitous banks, the rocks being in some places, bare and naked as a wall, and in others interspersed with rich forest timber. It is one continuous rapid the whole way, flowing with such impetuosity that at a point a little below the Whirlpool, where the channel is more than usually contracted, the level of the water in the middle is elevated from five to seven feet above that of the current at either side. But let me hurry to the Falls.

After taking some refreshment in Queenston, I proceeded by a private conveyance along the main road, preferring that to the railway, on which the trains are drawn by horses. Mounting the steep hill which rises directly from the town, I had ample opportunity of surveying the battle-ground on which was fought one of the sharpest conflicts in the annals of the war of 1812. The British were the victors on the occasion, and the monument raised to the honour of their commander, who fell gloriously on the field, occupied the highest point of the hill. It is as tall, and quite as ugly, as the Duke of York's column in Waterloo-place. A rent several inches in width traversed it from the pedestal to the capital, occasioned by an attempt made to blow it up with gunpowder, by a vagabond connected with the insurrection in 1838, whose ambition was on a level with that of the wretch who fired York Minster. On gaining the top of the hill, the road for a little distance wound very near the verge of the precipice, at a point where several of the American troops were driven over the crags during the conflict. Before proceeding any further I turned round to gaze on the prospect which spread beneath us. It was gorgeous and extensive. The level of Lake Ontario was displayed for a great distance on either hand to the view, large sections of Canada and New York, richly cultivated, lying, as it were, beneath our feet, the broad blue lake itself forming a glorious background to the picture.   From the top of the monument the view is still more extensive, Toronto being visible on the opposite side.

It was a warm still evening, and it was only after a brisk drive of nearly an hour's duration that I came within reach of the cataract's voice.   I had been long listening for its thundering tones, but could not distinguish them until I was within a couple of miles of the Falls.   Were Niagara calling aloud from a hill-top, there might be some foundation for the fabulous accounts which are sometimes given of the distances at which it can be heard; but thundering as it does at the bottom of a deep chasm, its mighty roar is smothered amongst the crags that rise around it.

I drove up to the Pavilion Hotel, situated on a high bank which overlooks the cataract.   A lovely moon was by this time shining in the deep blue sky, the air was rent with unceasing thunders, and the earth as I touched it seemed to tremble beneath my feet.

To my surprise and delight I found a large party of Canadian friends at the Pavilion.   They had but just arrived after a fatiguing journey from the West, and, with the exception of three, were preparing to retire for the night.   The three consisted of two ladies and a gentleman, who were determined to enjoy a moonlight view of the Falls.   It needed no very great persuasive powers to induce me to accompany them ; so after ordering a good supper to be prepared for us, we set out in search of the cataract.

The high, wooded bank on which the Pavilion rests, rises for nearly 200 feet above the upper level of the Niagara River.   It has consequently to be descended before the tourist finds himself upon the level of the verge of the cataract. From the observatory on the top of the Pavilion it is visible in all its length and depth, but from the windows and balconies of the hotel the American Fall only can be seen, the lofty trees on the bank screening the great Canada Fall. The moon being in the south, the face even of the American Fall, which has a north-western aspect, was buried in the deepest shade. We could hear the voice of the cataract in all its majesty, but as yet got no glimpse of its terrible countenance.

Passing through the garden behind the hotel, and emerging from a small postern gate, we found ourselves on the top of the bank. We had a guide with us, and needed him. Our path zigzagged down the steep descent, and we had to grope and feel our way, which was only occasionally visible to us by a few faint bars of moonlight falling upon it after struggling through the foliage. At last we got upon level ground, and as we threaded our way through the heavy timber, we became more and more enveloped in the spray. Emerging from the dense wood of the bank, we found ourselves, after a few steps in advance, upon TABLE ROCK.

Drenched and blinded as we were by the dense spray, which now fell less in showers than in masses around us, for a time we could see nothing, although a roar as of ten thousand thunders fell upon our ears. At length, after recovering ourselves, we looked in the direction of the cataract, but for a few minutes we could discern nought but the thick mist, in which we were enveloped, faintly illuminated by the moonbeams. A slight puff of wind at last drove it a little aside, and revealed to us the rapids above, gleaming in the cold moonlight as they shot and foamed over the rocky channel. We could thus trace them to the very line where the maddened waters took their great leap, beyond which all was darkness, mystery, noise and turmoil. We could observe the cataract take its plunge, but could not catch a single glimpse of its descent, or of the abyss into which it fell. In addition to the roar of the falling waters, a hissing noise stole up to us from the chasm, produced by the seething and foaming river beneath, whilst every now and then the faint voice of the American Fall, far below upon our left, would mingle with the deep chorus which swelled around us. We were within a few feet of the verge of the chasm where we stood, each having hold of the guide, who warned us not to approach a step beyond the spot to which he had led us. Although we saw nothing beyond the rapids above the Fall, the grey mist, and occasionally Goat Island, which loomed in spectral outline through it, there was something awful and sublime in the deep obscurity and the mystery which reigned over the scene, the impressiveness of which was enhanced by the incessant thunders which emanated from the abyss.

On returning to the hotel, I immediately mounted to the observatory, from which I enjoyed a magnificent prospect. Goat Island lay beneath me, as did also the American bank, and the branch of the river which rolled impetuously between them, as well as the whole of the rapids, between the island and the Canada shore. But from the verge of the cataract downwards, the moonbeams were absorbed by an enormous cloud of spray. When I retired to rest, notwithstanding all my efforts to get a sight of them, I had as yet only seen where the Falls were, but not the Falls themselves; but I consoled myself on going

with the American bank, where it strikes it, and giving the American Fall the appearance of being occasioned by a tributary here uniting with the main stream, and tumbling over its rocky and precipitous bank. The dry precipice of Goat Island occupies about a quarter of the whole extent of the ledge, one half of it being fully appropriated by the Canada, and the remaining quarter by the American, Fall. Notwithstanding the great height of the fall, which is from 170 to 200 feet, its enormous width gives it, when the whole is seen at a glance, the appearance of being wanting in altitude.

The reader who is acquainted with the localities of London, may, from the following illustration, form some faint idea of the magnitude of Niagara. Let him suppose a ledge of rock, nearly as lofty as its towers, commencing at Westminster Abbey, and after running down Whitehall, turning, at Charing Cross, into the Strand, and continuing on to Somerset House. Let him then suppose himself on Waterloo bridge, whence every point of the mighty precipice could be seen. Let him lastly suppose an immense volume of water falling over the whole of it, with the exception of a portion extending, say, from the Home Office to the Admiralty, which is left dry,—and he may have some notion of the extent of the great cataract. The tumbling and foaming mass extending from 'Somerset House to the Admiralty, would, with the bend at Charing Cross, occupy the place of the Horseshoe or Canada Fall; the dry rock, between the Admiralty and the Home Office, that of the precipice of Goat Island; and the continuation of the cataract, between the Home Office and the Abbey, that of the American Fall.

Notwithstanding the magnitude of its proportions, it must be confessed that the first sight of it disappoints the majority of those who visit it. The reason of this, in my opinion, is, that the first view of it is obtained from an elevation far above it. In attempting to picture it to themselves before seeing it, people generally place themselves in a position from which they look up to it. The lower level of a fall is decidedly the most advantageous point from which to view it ; and were Niagara first seen from below, the most magnificent creations of fancy would be found to come far short of the reality. But when, instead of being looked up to, it is looked down upon, one's preconceived notions of it are outraged, and the real picture is almost the inverse of the fancy one. Besides, to see it all at a glance, you must stand a considerable distance from it, and the angle with which it then falls upon the eye is much smaller than if you attempted to grasp it from a nearer point of view. But, despite the first disappointment, no one remains long enough about Niagara to become familiar with it, without feeling that the reality is far grander and more stupendous than he had ever conceived it to be. Such was the case with myself. I have visited Niagara four different times, my average stay each time being about five days, and left it each time more and more impressed with its magnitude and sublimity. At first one regards it as a whole, of the extent of which he can form no very definite idea ; but, by-and-by, he learns to estimate its magnitude, by applying to it appreciable standards of measurement. When he comes thus to understand it, he finds that the American Fall, the smaller of the two, would of itself have sufficed to meet all his preconceptions.

No one should stay for less than a week at Niagara. There are scores of different points from which, to appreciate it, it must be viewed. It should be seen from above and from below the point at which it occurs; from the level of the ledge from which it plunges, and of the abyss into which it falls; from the top of the bank far above the rapids, and from the boiling and surging ferry, over which the tourist is conveyed by a small boat almost to the foot of the American Fall. It is when viewed from the top of the American bank close to this fall, that its enormous width can be best appreciated. It should also be seen from every point of Goat Island from which a view of it can be obtained. The island is gained by a wooden bridge, which crosses the American branch of the river in the very midst of the rapids. How a bridge could be constructed on such a spot, baffles comprehension. On your left as you cross, such is the rapid descent of the channel, that the water seems to pour down the side of a hill. On your right is the verge of the American Fall, not a furlong off. You are conscious that, should you fall in, a single minute would suffice to plunge you into the abyss.

Once on Goat Island, you are between the cataracts, both of which you may see from different portions of its wooded surface, as well as from the bottom of its precipice, which you can descend by a spiral wooden staircaise. When you descend, you are still between the cataracts, being now, however, at their feet, instead of on their upper level. To get from the one to the other, you have to scramble over broken masses of rock, and along narrow ledges which have been converted into pathways. Let not the tourist forget to place himself close to the American Fall on the upper level of Goat Island. If the day is bright, and he has an eye for colours, he will linger long to enjoy the rich treat before him. Taking a mere casual glance at it, the falling mass appears to be snow-white, but by looking steadily into it he can analyse the white into almost every colour and shade. This he can also do on looking at the Horseshoe Fall from the other side of Goat Island. It is from this point that the rainbow which spans the chasm, when the day is bright, is best seen. You have to look far down upon it, for it lives only amid the snow-white spray which mantles the foot of the cataract.

On the Canada side of the Horseshoe Fall, the tourist can pass for about 150 feet between the sheet of water and the rock. Whilst there, he perceives how the cataract is gradually receding. The rock below crumbles before the action of the water, and the superincumbent mass falls when it is deprived of sufficient support underneath. The rate at which it thus recedes is about a foot per year. At this rate it must have taken about 4,000 years to wear its way back from Queenston. It is still about eighteen miles from Lake Erie, which, at the same rate, it will take upwards of 100,000 years more to reach ! It is worth while to go under the sheet, were it only for the view of the fall which you obtain from the foot of the spiral staircase by which you descend from Table Rock. There is no other spot from which Niagara can be seen in all its majesty as it can from this. You are close to the great fall, and at its very feet. Looking up to it you see nothing but it and the heavens above it, when it appears like a world of waters tumbling from the very clouds. At its two extremities the water is of a dazzling white, from the point at which it takes its leap ; but in the centre, and in the deepest part of the bend, where the volume is greatest, it preserves its pale green colour, streaked with white veins like marble for fully two-thirds of its way down. Let me repeat, that but for this view it would not be worth while to go under the sheet ; to do which, one has to change his warm dry clothing for a cold wet oil-cloth suit, and his boots for heavy clogs which are soaked from morning till night, and to penetrate masses of eddying spray which nearly blind and choke him, under the guidance of a damp negro, who is never dry.

Niagara would appear to greater advantage were its adjuncts on a much greater scale. It is like a vast picture in a meagre frame. The banks are lofty, picturesque, and bold ; but they are by no means on a scale commensurate with the magnitude of the cataract. It has no rival in the admiration of those who behold it. It is itself the only object seen or thought of, when you are in its presence.

The walks about Niagara, along both banks, and on Goat Island, are numerous and attractive, those on the island particularly so.

Once seen, the impression which it leaves is an enduring one. It becomes henceforth a part of one's intellectual being, not the plaything of his imagination, but the companion of his thoughts. You can recall at pleasure every feeling and emotion which it conjured up on first beholding it. As I saw Niagara and heard it then, so I see and hear it now.

CHAPTER V.

ARTIFICIAL IRRIGATION OF THE UNITED STATES.—
RIVALRY BETWEEN CANADA AND NEW YORK FOR
THE CARRYING TRADE OF THE NORTH-WEST.—THE
NAVIGATION LAWS.

Buffalo.—The Canal System of the United States.—The Erie Canal.
—Other New York Canals.—The great and subsidiary Canals of
Pennsylvania.—The Chesapeake and Ohio Canal.—The James
River, and Kanawha Canal.—Canals in the Mississippi Valley.—
The Carrying-trade of the North-west.—What it is.—The Region
constituting the North-west.—The Lakes, and the Lake-trade.—
Comparison of the Routes to Tide-water, from the foot of Lake
Erie, through New York, by the Erie Canal, and through Canada,
by the St. Lawrence and the St. Lawrence Canals.—The injurious
effects of the Navigation Laws upon the Trade of Canada.—The
necessity for their Repeal.

ON leaving the Falls I ascended the river to Buffalo,
which is situated at the foot of Lake Erie, on the
American bank, and is at present the great *entrepôt* of
the internal trade of the North-west. Although
nearly 600 miles from the coast, Buffalo exhibits all
the characteristics of a maritime town. Indeed, to
all intents and purposes it is so, the lake navigation
with which it is connected being in length at least
equal to that of the Mediterranean. It is beautifully
situated on a sloping bank overlooking the lake, and
is built and laid out with all the taste which marks in
this respect most of the towns of Western New York.
Its population is about equal to that of Rochester;

and the two towns keep abreast of each other in their rapid progress. Buffalo has more of a floating population than Rochester,—a feature in which it resembles all towns which partake more or less of a seaport character.

Lake Erie is preeminently an American, as Lake Ontario is a Canadian, lake. Both serve equally as the boundary between the British province and the Union ; but Lake Ontario is far more of a highway for Canada than it is for New York, whilst Lake Erie is less so for Canada than it is for Ohio, Michigan, and the territories which lie beyond. It is British trade that predominates on the one, whilst American traffic has no rival on the other. It depends upon circumstances, which will be presently alluded to, whether this distinction will permanently prevail. Much lies in our power in reference to it ; and the policy of England may yet render lake Ontario as much a highway for the great North-west as Lake Erie has hitherto almost exclusively been.

Buffalo occupies the same position as regards Lake Erie, and the great artificial artery of New York, as Kingston does in regard to Lake Ontario and the line of public works which extends from it to tide-water in the St. Lawrence. Both are the points at which the great natural channels of communication are abandoned in descending to the coast, and the rival artificial means of transport are resorted to, which the energy and enterprise of the State and the Province have conjured into existence. At Buffalo the lake navigation terminates for such goods as are destined for New York, whilst at Kingston it ends for such as are on their way to the Ocean by Montreal and Quebec. But before considering the respective claims of the rival routes, it may be as well here to take a brief glance at what has been done in the United States for the improvement of the country and the furtherance of trade, by the construction of canals. A better spot from which to contemplate the artificial irrigation of the Union could not be chosen than Buffalo, which is at the western extremity of the great Erie Canal.

What has been said in a previous chapter on this subject renders it unnecessary to dwell at any great length upon it here. The canal system in America resembles very closely in its distribution the Railway system already considered. Its chief design is to connect at the most favourable points, the great sections of the continent separated from each other by obstacles which they have been constructed to obviate. The New England canals have less of a general than a local importance; whilst many of those which lie to the west and south constitute the most practicable media of communication between vast sections of the Confederacy, which would otherwise, for the purposes of heavy traffic at least, be virtually isolated from each other. There is no great coast system of canals, resembling the coast system of railways; for the obvious reason, that the Atlantic, which may not be the best highway for travellers, furnishes a better means for the transport of goods from one point to another of the coast region than any line of canals would do. Most of the great canals, which have a national importance, unite the coast region either with the basin of the St. Lawrence, in the neighbourhood of the Great Lakes, or with the valley of the Mississippi. The chief of these are the Erie canal, the Pennsylvania canal, the

Chesapeake and Ohio canal, and the James River and Kanawha canal.

One of the oldest canals, and decidedly the first in point of importance in the Union, is the Erie Canal, which unites the sea-board at New York with the region bordering the lakes at Buffalo. Its eastern extremity is at Albany, where it joins the Hudson 160 miles above New York. Thence it proceeds westward along the valley of the Mohawk, passing through or by the towns of Schenectady, Canojoharie, Little Falls, Utica, Rome, Syracuse, and Palmyra, crossing the Genesee valley at Rochester, and then proceeding westward by Lockport to Buffalo. Its entire length is about 370 miles. It puts the city of New York in connexion with the lakes, far above the rapids of the St. Lawrence, and beyond the Falls of Niagara; in other words, it opens up to New York the trade and traffic of the vast basin in which the lakes lie. It also connects at two different points, though by a circuitous route, the sea-board with the Mississippi valley; the Genesee valley canal uniting it, as already seen, at Rochester with the Allegany River in Pennsylvania, which is one of the parents of the Ohio; and the great Ohio canal connecting Lake Erie, in which the Erie canal terminates, with the Ohio River.

This majestic work was planned and executed by De Witt Clinton, Governor of New York in 1817 and 1822. He met with every opposition in carrying out the work. Numbers consoled themselves with the reflection that if they lived to see the "Clinton ditch," as it was contemptuously termed, finished, they would indeed attain a green old age. Others laboured to convince their fellow-citizens that in a financial point of view the scheme would prove an utter failure. But the great heart of Clinton was not to be daunted, either by ridicule or by more sober opposition, and he persevered with his plan, staking his political reputation upon its success. The whole of the western portion of the State was then a wilderness, and he was advised to postpone the undertaking until that section of it had advanced somewhat in the career of material improvement. But he declined listening to such advice, determined that his canal should be the great agent in improving the West—the cause, not the consequence of its advancement. The nature of the country favoured his scheme. In two different places sections of the canal could be constructed for seventy consecutive miles without a lock. He commenced his operations on the more easterly of these great levels, and after finishing that link of his work, had detached links of it constructed elsewhere along the intended line. These were found to be so useful in their different localities, that the whole community soon became clamorous for the completion of the undertaking; and thus before it was finished, the governor, who at the commencement had stood almost alone, had the vast majority of his fellow-citizens with him. At length the great work was completed, and a salute, which was fired from guns placed along its bank, at regular intervals, the whole way from Buffalo to Albany, announced that it was opened throughout the whole line. The result justified the expectations of the sagacious and adventurous Clinton. In the course of a few years an almost miraculous change was effected in the whole aspect of Western New York. The forest suddenly disappeared, and towns and villages sprang up on sites which had long been the haunts of the savage, the wolf, and the bear. About twenty-four years ago, the forest completely shrouded the region, which is now the granary of the State. In addition to the benefit which it thus conferred upon New York, it gave a stimulus to the settlement of the vast tracts to the westward ; leading to the sub-jugation, by civilized man, of the wilds of Ohio and Michigan.

In its original dimensions the canal was forty feet wide and four feet deep. Such, however, has been the increase of the traffic upon it, that, in order to accommodate it, the canal has of late years been quadrupled in size; that is to say, it has been made eighty feet wide and eight feet deep. This enlarge-ment, which involved a much greater outlay than that required for the original construction of the canal, is not yet completed throughout, but is steadily progressing. The success of the Erie canal and the revenues derived from it, enabled the State to em-bark upon other projects of a similar character, but of minor extent ; and thus have arisen those numerous lateral canals with which the tourist so frequently meets in the State. The success of the New York system, of which the Erie canal is the great feature, is indicated by the surplus revenues now derived from it. The aggregate cost of the canals was about thirty-one millions of dollars, the average interest payable upon which is five-and-a-half per cent. The net revenue from all the State canals, after deducting the cost of collection and of superintendence, is upwards of two millions. This is nearly equal to seven per cent. upon the whole cost, or one-and-a-half per cent. beyond the average rate of interest payable upon it.

The new Constitution adopted by New York in 1846, provides for the establishment of a sinking-fund, for the extinction of the debt; the object being to pay both principal and interest in the course of about twenty years. The annual revenues of the canals will then be available for further improvements, or for defraying the expenses of civil government in the State. As these expenses do not exceed a million and a quarter dollars per annum, and as the net receipts from the canals will be at least three millions per annum twenty years hence, it is obvious that, in these magnificent works, the New Yorkers have not only a means of ultimately ridding themselves of taxation for the support of their government, but of carrying on the work of internal improvement to an almost indefinite extent.

The Pennsylvanians, in constructing their great canal, which pursues a line almost parallel to that of the Erie canal, had a double object in view—that of facilitating their own internal, particularly their mineral, trade, and of creating a rival to the New York canal within the limits of Pennsylvania. It interferes but little, however, with the traffic of the Erie canal; a result which has not a little contributed to plunge Pennsylvania into those financial embarrassments, out of which she now begins to see her way. Her great western line of canal is of immense service to her own internal trade, and being the most northerly of the canals which unite the sea-board directly with the Mississippi valley, will yet play an important part in the conduct of the trade between them. As is the case with its great rival, the Pennsylvania canal is the chief trunk line of communication from one extremity of the State to the other, many lateral branches, or tributary canals, leading into it. There are several other improvements of this kind in Pennsylvania which have no direct connexion with its great canal system.

The next great public work of the kind with which we meet is the Chesapeake and Ohio canal, extending along the borders of Maryland and Virginia, but being chiefly, if not exclusively, within the limits of the former. It commences at Alexandria, in the State of Virginia, where it unites with the Potomac; and, after ascending the Virginia bank of the river for about seven miles, crosses it by a stupendous wooden aqueduct, on stone piers, at Georgetown, after which it ascends the valley of the Potomac on the Maryland side. The object of this canal is to unite Chesapeake Bay with the valley of the Ohio; in other words, to open a direct communication further south than Pennsylvania, between the great valley and the sea-board. The project, however, is not yet completed, the canal having only obtained about 180 miles of its intended length. It will be some time ere the remainder is constructed, but the necessities of the growing trade of the country on both sides of the Alleganies forbid the notion of its continuing very long in its present incomplete state.

This canal has much more of a national importance attached to it than the next and most southerly of the great lines of improvement, designed to facilitate the access of the products of the great valley to the ocean. The James River and Kanawha canal, which is also unfinished, is designed to unite the river at Richmond with one of the navigable tributaries of the Ohio. This, when completed, will constitute another link of connexion between the sea-board and the Mississippi valley. It would appear, however, that this canal is destined to have more of a local than a great sectional importance annexed to it. It will be of immense advantage to the State of Virginia, in which it lies, particularly to the central valley, and the portion of the State west of the Alleganies, to which markets were formerly very difficult of access. But the chief trade between the sea-board and the northern section of the Mississippi valley, will be carried on by means of the more northerly lines of communication, passing through Maryland, Pennsylvania, and New York. There are numerous canals in the Southern States, but none on the same magnificent scale as those just alluded to. Their chief object is to unite the great navigable streams which irrigate the eastern section of the valley, some of which fall into the Gulf of Mexico, and others into the Ohio. The very skeleton of the canal system which will one day irrigate the valley on both sides of the Mississippi, is not yet formed. The natural irrigation of the region is on so magnificent a scale, that it will be long ere its increasing population occupy all the banks of its great streams. Until that is done, and multitudes of industrious people are settled back from the rivers, with the exception of a few of obvious utility to large sections of the country, canals will not multiply with great rapidity in the valley. Those which descend into it through Ohio, Indiana, and Illinois, are not of mere local importance, their object being similar to that of the eastern and western lines, passing through New York, Pennsylvania, Maryland, and Virginia, to unite one great section of the Union with another; in other words, to connect the valley with the region of the lakes, and, consequently, with the sea-board.

Such is the scope taken by the canal system of the United States as already developed. It has been laid out upon a scale which will enable it to meet the wants of double the present population of the country. It will not be long, however, ere it undergoes considerable expansion, that it may meet anticipated exigencies; for many of the present generation will live to see the population of America trebled, if not quadrupled. The feature in this gigantic system most interesting to us is decidedly that which it exhibits in the State of New York; not only from the contiguity of that State to our own Provinces, but also from the rivalry which exists between it and Canada for the carrying-trade of the North-west. As this is a struggle the issue of which will be materially affected by the continuance or the abrogation of the navigation laws as regards the St. Lawrence, I cannot here do better than devote the remainder of this chapter to an explanation of it.

In doing so, let me first describe what the carrying-trade of the North-west is. The region known in America as the North-west, comprehends not only the whole of that portion of the United States territory lying in the basin of the St. Lawrence, west of the lower end of Lake Erie, but also a considerable section of the northern side of the valley of the Ohio, and the upper portion of that of the Mississippi. It thus embraces an enormous area, comprising a small part of Pennsylvania, the greater portion of Ohio, the whole of Michigan, the greater parts of Indiana and Illinois, the whole of Wisconsin, and nearly all Iowa. In other words, it includes nearly the whole of six of the States of the Union, lying south and west of the lakes, as well as, for the purposes of this inquiry, all that portion of Western Canada which lies upon Lake Erie, and which constitutes the eastern and northern shores of Lake Huron, and the north coast of Lake Superior. The whole of this immense expanse of country, with the exception of the part of Canada lying to the north of the two last-mentioned lakes, is fertile and arable, comprehending indeed the finest grain-growing districts, not only in Canada, but also in the United States. To this may be added, as being involved in the question, the greater portion of Western New York; in other words, the granary of that important State. The population of this enormous region is at present between five and six millions, being the most active and enterprising of the inhabitants of the continent. It is here, too, that the population of the country is increasing at the most rapid ratio. Some notion may be formed of the rate of its increase, when it is known that, during the five years ending in 1845, Illinois added about forty-five per cent. to the number of its people, whilst the population of Michigan during the same period was increased by nearly fifty per cent. In 1850, the population of each of these States will be double what it was in 1840; notwithstanding the stream of emigration which has latterly set in for California and the Pacific.

Partially peopled and partially cultivated as it yet is, the trade of this great region has already attained a gigantic expansion. It is almost exclusively agricultural, and it is to agriculture that it will mainly look as the source of its future wealth. Its surplus products will procure for it from other quarters the necessaries and the luxuries of life. Its annual surplus is already great, and is being constantly exchanged, either in New England or in foreign markets, for such commodities as its increasing millions may be in want of. The conveyance of its produce to the sea-board, and the transport into the interior from the coast of such articles as it receives for consumption in exchange for them, constitutes what is called the carrying-trade of the North-west. It is to lead this trade through its own territory that Canada is now competing with New York to become the forwarder, at present of five millions, and prospectively of fifty millions of people.

It is not only in the vast extent of the region contemplated, in the fertility and varied capabilities of its soil, and in the unquestioned enterprise of its inhabitants, that consist the elements of a great trade. Situated far in the interior, nature has been lavish in the advantages she has conferred upon it. By means of the great lakes, their tributaries and connecting links, it can not only carry on an extensive trade within itself, but also approach from the interior of the continent to within 300 miles of tide-water. The facilities thus thrown in its way for trading, not only with itself, but with the foreign world, has a manifest tendency still more rapidly to develope its resources and extend the limits of its wants. The lake trade of the United States, comprising that carried on with Canada, is second only in importance to that of the sea-board itself. Had it been necessary to provide it throughout its length and breadth, with artificial channels of communication with the coast, it would have been long ere the means could have been secured to meet so enormous an outlay. But its great inland seas not only enable it to transport its produce at comparatively little cost, to a point not far from the ocean, but also afford it all the facilities for traffic which an extensive coast implies. With the single exception of Iowa, there is not one of the States named but has a lake coast, of more or less extent, each having its own harbours and lake trade. The aggregate lake coast which the North-west possesses, taking into account only that of the five principal lakes, constituting the great freshwater chain of the continent, extends for upwards of 4,000 miles. This is more than the whole circumference of Great Britain and that of Ireland in addition. The lake coast of Canada alone, from the western extremity of Lake Superior to the eastern limit of Lake Ontario, extends in one continuous line for nearly 2,000 miles. The great inland highway which nature has bountifully provided for it is thus accessible to almost every portion of this highly favoured region, stimulating its occupants to additional activity from the facilities which it affords them for disposing of their produce. The simple question between Canada and New York is, which can best supply the link wanted to connect the North-west with the ocean.

A glance at the map will suffice to show that, for a considerable time to come, the lakes, as far down as the foot of Lake Erie, will form the common high-way for all parties inhabiting their shores. When population greatly increases along the Canada coast of Lake Huron, the northern portion of Michigan, the shores of Lake Superior, and for some distance down both banks of Lake Michigan, it is not improbable that the exigencies of the transport trade of these regions will then lead to the opening of a canal communication between the Georgian Bay on Lake Huron, through Lake Simcoe to Lake Ontario, in the neighbourhood of Toronto. The whole distance from Penetanguishine on Georgian Bay, to Toronto on Lake Ontario, is not over ninety miles, Lake Simcoe, which is thirty-five miles long, lying in the direct line between them. This would limit the cutting to about fifty-five miles, nearly forty miles of land intervening between Toronto and the southern end of Lake Simcoe, and but little more than fifteen dividing its northern extremity from the great arm of Lake Huron alluded to. Lake Ontario will then be accessible to a large portion of the North-west, without resorting to the circuitous navigation of the southern section of Lake Huron, the River and Lake St. Clair, and Lake Erie. But for the present we may regard the lakes to the foot of Lake Erie as the common and most practicable highway for the whole region, the foot of Lake Erie being, therefore, the point upon which, for a long time to come, its accumulated products will be poured. This will always be the case as regards the products of the great bulk of Michigan, the northern section of Ohio, a portion of Pennsylvania and Western New York, and a large part of the best portion of Upper Canada. It is here then—at the foot of Lake Erie—that the rival routes to the ocean commence.

I have already described the means provided by the State of New York for continuing, through its own territory, the transport-trade of the lakes. The produce transported from the interior to Buffalo, may thence be conveyed by the Erie canal to Albany on the Hudson, by which it can descend to New York. But before considering the respective merits of the rival routes, it may be as well, that through New

York having been described, to give an account of
the route through Canada.

Whilst it is the object of the New Yorker to make
his chief river, the Hudson, available for his purpose,
that of the Canadian is to do the same by the St.
Lawrence.   Had the navigation of this noble stream
been continuous from the Lakes to the ocean, the
struggle, if it would ever have arisen, would not have
been of long duration.   But it meets with frequent
and formidable interruptions, to the removal of which
the government and people of the province have
applied themselves with vigour, perseverance, and
complete success.

The first interruption to the navigation of the St.
Lawrence downwards occurs a little below Lake Erie,
developing itself in the formidable character of the
Falls of Niagara.   To obviate this difficulty, the
Welland canal has been constructed through the rich
agricultural district intervening between Lake Erie
and Lake Ontario.   The canal starts from a point on
the Canada shore of Lake Erie, a considerable distance above Buffalo, and enters Lake Ontario at St.
Catherine's, a little west of the mouth of the Niagara
River.   From the latter point, nearly the whole length
of Lake Ontario lies in the direct route to the ocean.
At the foot of the lake is the town of Kingston,
from which the St. Lawrence is navigable through
the " Thousand Islands " to Brockville, and thence
to " Dickenson's Landing," about 120 miles below
Kingston.   There are several rapids between the
last-mentioned place and Prescott about thirty miles
above it, but they are not of a character sufficiently
formidable to constitute any serious impediment to

the navigation of the river. At Dickenson's Landing is the first and most stupendous of the series of rapids which intervene between it and Montreal. This great obstruction, which is upwards of twelve miles in length, is avoided by means of the St. Lawrence canal, extending along the north bank of the river, from Dickenson's Landing to Cornwall at the head of Lake St. Francis. The next interruption arises from the rapids which occur between Lake St. Francis and Lake St. Louis. To avoid it, the Beauharnois canal has been constructed on the southern bank of the river, forming a practicable link of communication between the two lakes. At the foot of Lake St. Louis the last great obstruction is encountered in the shape of the formidable rapids of Lachine, which are avoided by means of the Lachine canal, uniting the lake with the St. Lawrence at Montreal, immediately below the rapids. Another impediment, but of a less formidable character, is met with somewhat lower down in Lake St. Peter, the volume of which is very shallow and the channel frequently shifting. To obviate the latter difficulty, works have been constructed in connexion with some of the numerous islands at its upper end, the object of which is to straighten the channel and to render its position permanent. At Three Rivers, a few miles below Lake St. Peter, but upwards of 450 from the mouth of the river, tide-water is reached, beyond which the channel of the St. Lawrence is practicable to the Gulf. The portion of the river interrupted by rapids, and extending from Dickenson's Landing to Montreal, is, including the two lakes St. Francis and St. Louis, a little upwards of one hundred miles in length, but exclusive of the lakes only about thirty, which is also about the aggregate length of the canals by which they are avoided. There is still another route from Kingston to Montreal by way of the Rideau canal, which extends from the foot of Lake Ontario to Bytown on the Ottawa, this river uniting with the St. Lawrence at the head of Lake St. Louis. But this route is both more tedious and expensive than the descent by the St. Lawrence, and is more adapted for military and local purposes than for constituting an eligible link in the chain of communication between the North-west and the sea-board.

Such, then, being the rival routes through New York and Canada, let us now consider the advantages which they respectively offer, in the transport of produce to the sea-board, and of imports to the interior. The question between them turns upon the saving of time and expense. The route which can accomplish its object at the least sacrifice of both, will carry all before it ; whereas if one has the advantage only in point of time, and the other only in point of expense, the issue may remain doubtful for some time yet to come, unless the advantage possessed by the one be so great as to neutralize that enjoyed by the other. There can be no better mode of showing how the case stands between them than by following a cargo of produce from the interior to the sea-board, first by the one route and then by the other, noting, in either case, the time consumed and the expense incurred on the way. Let us in the first place take the route of the Erie canal.

We have already seen that, as regards the rival routes, the lake navigation terminates at the foot of Lake Erie. The produce conveyed to Buffalo from either shore of that lake, or from the regions bordering the lakes above it, is carried thither either in sloops, schooners, or steamers. The last mentioned are generally employed in the transport of passengers and the lighter kinds of goods, the great bulk of the produce which descends the lakes being conveyed by means of sailing vessels. On reaching buffalo the cargo must be transshipped, the navigation of the Erie canal being confined to boats built for the purpose. This, supposing the schooner to be of 300 tons burden, will occupy at least two days. The cargo, being distributed into different boats, then proceeds on its tedious canal journey of nearly 400 miles in length. Three miles per hour is the maximum rate of speed authorized by law for freight boats, a greater speed being unattainable without injury to the canal banks. Making proper allowance for stoppages at locks, and for other detentions by the way, the average speed along the entire length of the canal will not exceed two miles per hour. Taking the canal as 375 miles long, and supposing the boats to continue moving at this rate day and night without intermission, they would occupy seven days and nineteen hours in reaching Albany, say eight days, which, with the two consumed at Buffalo, make ten days as the shortest time in which a cargo can reach the Hudson after arriving at the foot of Lake Erie. As the canal boats do not navigate the Hudson, another transshipment takes place at Albany, into barges constructed to descend the river. This will occupy at least another day, whilst the greater part of two days more will elapse ere it reaches New York. Here, then, we have thirteen days consumed at the least, in the transport of a cargo from Buffalo by the Erie canal and the Hudson to New York. So much for time—now for expense. The first item of expense incurred is for transshipment at Buffalo. This upon a barrel of flour or a bushel of wheat may be but trifling, but it is of trifles that the largest aggregates are made up. Then comes the cost of transport along the canal, which is materially enhanced by the heavy canal dues which have to be paid. The ordinary rate at which a barrel of flour and a bushel of wheat can be conveyed from Buffalo to Albany is 2s. 7d. sterling, and 9d. sterling respectively. There is then the cost of transshipment at Albany, and the freight to New York, which is rather heavy, inasmuch as the barges which descend the river have to be towed by steam. The entire cost from Buffalo to New York, including all charges, may be taken at 2s. 9d. sterling for a barrel of flour, and 10d. sterling for a bushel of wheat. Such is the sacrifice both as to time and money, at which a cargo, descending by this route to the sea-board, is brought to the point from which it starts on its ocean voyage. Let us see how the case stands with regard to the St. Lawrence.

We are once more at the foot of Lake Erie, on board a schooner propelled by a screw, laden with produce from the upper country. But we now take the route to the left, instead of that to the right as before, and at once enter the Welland canal.

This is the proper place to mention the essential difference which exists between the internal improvements of Canada and New York. The Erie canal is unsurpassed in length, but even on its enlarged scale it is small, both in width and depth, as compared with the Canadian canals. These, as already shown, are exceedingly short, occurring at intervals; and as their design is to render continuous the navigation of a vast river, they are on a scale, as to their other proportions, commensurate with their object. They are, in fact, ship canals. This has an important bearing upon the question at issue between the parties. It renders unnecessary, in pursuing the Canada route, the delay and expense of a double transshipment, such as I have shown must necessarily take place at Buffalo and Albany. The consequence is that the vessel which descends to the foot of Lake Erie with produce, can pursue her journey by the Canada line, either to Montreal or Quebec, without once breaking bulk. Much of the traffic by this line is already carried on by screw propellers, some of which are upwards of 300 tons burden; and there is little doubt that ere long they will entirely supersede sailing craft, in the direct transit trade by the Canadian waters between the interior and tide-water.

Having emerged into Lake Ontario from the Welland canal, the propeller proceeds down the Lake to Kingston, whence she descends the St. Lawrence to Dickenson's Landing, at which point she takes the St. Lawrence canal to Cornwall, from which she descends Lake St. Francis to the Beauharnois canal, through which she passes into Lake St. Louis, at the foot of which she proceeds by the Lachine canal to Montreal, from which she can descend without impediment to Quebec. The whole distance from the foot of Lake Erie to Quebec is not over 650 miles, which a good propeller can accomplish, if well managed, in four days. It thus takes but four days to bring a cargo from the foot of Lake Erie, by the Canada route, to the point from which it starts upon its ocean voyage. In point of time, therefore, the Canada route has the advantage by no less than nine days over its rival.

The cost at which a cargo is forwarded at present

by this route, is no criterion by which to judge of what it will be when all the capabilities of the line are fairly developed. It now costs 2s. 4d. sterling to forward a barrel of flour, and 9d. a bushel of wheat, from Lake Erie to Quebec. But at present, for want of a sufficient number of propellers, much of the produce that descends Lake Ontario is conveyed to Kingston by steamer, where it is transshipped, to be forwarded to Montreal. This, of course, increases the expense—an increase which will be avoided when the propeller becomes the chief medium of transport on the line. Besides, from its very nature, the carrying-trade by the Canada route is at present, or has been until very lately, in the hands of a few wealthy capitalists. It is now being diffused over a larger number of competitors, which will occasion a still further decrease of cost. When all the appliances of the route are fairly brought to bear, it is not too much to expect that a barrel of flour can be conveyed from Lake Erie to Quebec for 1s. 6d. and a bushel of wheat for 7d. In point of cost, therefore, the advantage is, or will be, with the Canada route to the extent of 1s. 3d. per barrel, and 3d. per bushel. Thus, both as regards time and expense, it is superior, between the lakes and tide-water, to the rival route.

But the object of bringing the produce of the interior to tide-water, in either case, is not to leave it there, but to forward it still further on. In carrying out the comparison between the two lines of transport, let us suppose that Liverpool is the destination of the cargo. It is obvious that the decision of the question between them depends upon the advantages offered, in either case, by the whole route, and not merely by a portion of it. The facilities which one part of a line may present, may be more than counterbalanced by the impediments which clog it in another, just as the difficulties in the way of one part may be completely neutralized by the facilities of another. The two cargoes, the course of which we have followed, are now, the one at New York, and the other at Quebec. We have seen that, in the race to these two points, the Canadian has, in every respect, outdistanced his competitor. But the produce on his hands at Quebec has still to descend the St. Lawrence, for about 350 miles to the Gulf, which again it has to cross ere it enters upon the open sea, from between Newfoundland and Cape Breton. The cargo shipped at New York, on the other hand, is launched at once upon the open sea on its way to its destination. There can be no doubt that, as regards this, the latter portion of the two routes, the natural advantages are with the New Yorker. But the question is, Do these advantages so greatly preponderate in his favour between port and port, as to counterbalance the disadvantages under which, as compared with his rival, he labours throughout the overland portion of the route ?

In considering this branch of the subject, we find that it is the misfortune of the Canadian to have to combat, on proceeding from port to port, not only with difficulties of a natural kind, but with others of artificial creation. He has not only the lower portion of his river and the Gulf beyond it to traverse ere he gains the open sea, but his movements are clogged with imperial restrictions, which fetter him in the form of navigation laws. Just at the point at which his triumph over his greatest obstacles is complete, and when he is called upon to contend with

H 3

some remaining difficulties of an ineradicable character, he finds his further progress impeded, not by
natural obstructions, but by acts of parliament.   It
thus appears, that it is where the advantages of the
Canada route end, that those of the American begin ;
or, to view the case from the other side, that the disadvantages of the Canadian route commence precisely where those of the American terminate—at
tide-water.   In the race hitherto we have seen the
Canadian by far the more agile of the two—an advantage of but little avail to him so long as, for the
rest of the course, his feet are heavily fettered.   Let
us examine into the difficulties which beset him from
tide-water, with a view to ascertain how far they are
natural and insurmountable, and how far artificial,
and therefore removable.

Though starting from different points, vessels from
both ports, by the time they have accomplished about
one-third of their respective voyages, fall into almost
the same line in prosecuting the remaining two-thirds.
The point at which they thus fall into a common
course is in the neighbourhood of Cape Race, the
south-eastern angle of Newfoundland.   At this common point of departure, the competition between the
two routes, in point of advantage, terminates the
natural difficulties with which the Canadian has to
struggle, lying between Quebec and Cape Race.
The great advantage which the American possesses
is that, in making this point, he can avail himself
of the open sea the whole way ; whereas for five-
sixths of the way to it, from Quebec, the Canadian
is confined to his river and the Gulf.   Although the
line is a little circuitous, the distance from Quebec
to Cape Race is considerably shorter than that from

New York to Cape Race. But this advantage is neutralized by the delays which frequently occur in the navigation of the river. Unless the wind is favourable, a vessel ascending or descending the St. Lawrence has to drop anchor with every adverse turn of the tide. But, with a fair wind, there is no reason—there being good sea-room the whole way, for the channel of the St. Lawrence from Quebec to the Gulf is, on an average, from fifteen to twenty miles wide—why a vessel from that port should not make Cape Race in five days. It is only under the same propitious circumstances that a ship from New York can gain the same point; the chief difference between the two routes consisting in this, that, circumstances more frequently favouring it, a ship proceeding by the latter does generally make Cape Race in less time than one descending the St. Lawrence. But, in point of time, we have already seen that the Canadian has a gain at tide-water of fully nine days over his competitor. If, therefore, he took fourteen days to gain Cape Race, whilst the American only took five, it would but put the two parties on an equality with each other so far as time was concerned. But, in general, a ship descending the St. Lawrence does not take fourteen days to gain this point. It will be making ample allowance for the difficulties of the route, if we assign a vessel ten days as the average time required to reach it. This is double the time in which, under favouring circumstances, it can be reached from New York. This still leaves a balance of four days in favour of the Canadian route from Lake Erie to Liverpool.

Another natural obstacle in the way of the Canadian is that, for six months in the year, the St. Lawrence is impracticable, on account of the ice with which its channel is blocked up. But the same may be said of the Erie canal, not that its channel is blocked up with ice, but that for nearly five months in the year it is without water. It is not prudent to remain so long in the St. Lawrence, but a vessel may safely leave it as late as the 7th or 10th of November. About the beginning of May it is once more practicable, and vessels from Europe frequently arrive at Quebec during the first week of that month. To preserve the banks from the injury which would be effected by ice, the Erie canal is drained in the month of November, and is not filled again with water until April. There is thus not more than a month's difference between the time for which the St. Lawrence and that for which the canal is impracticable. In both cases, the chief transport business of the year must be condensed within the time for which the routes are capable of being used.

But the chief obstacle in the way of the Canadian, after reaching tide-water, is that which is of artificial creation. We have seen that, as regards time in transporting produce from Lake Erie to Liverpool, if the balance of advantages is not actually with him, it need not be against him. The same cannot be said with regard to cost, for in this respect, under existing circumstances, the American has on the whole route the decided advantage. The ground gained, in point of cheapness, by the Canadian between Lake Erie and Quebec, is more than lost by him between Quebec and Liverpool. Various reasons contribute to this, one of which is, that the navigation of the Gulf being at some seasons rather precarious, the rates of insurance on sea-going vessels and cargoes proceeding by the St. Lawrence are considerably higher than on those crossing the Atlantic from New York. But the chief reason is to be found in the high rate of freight charged between Quebec and Liverpool by those who monopolize the navigation of the river. The whole trade of the St. Lawrence is confined, by the navigation laws, to the British ship-owner, from which accrues the double disadvantage of exorbitantly high freights, and delay in the transport of produce to its destination. It frequently happens that the quays both of Montreal and Quebec are overladen with produce waiting for exportation, but which remains sometimes for weeks on the open wharves for want of sufficient tonnage to convey it to Europe. It not only thus incurs the risk of damage, but has to pay for its transport almost any price that the ship-owner chooses to impose. So great is the disparity in this respect between Montreal and New York, that I have known 7s. 6d. sterling asked at the former for every barrel of flour to be conveyed to Liverpool, whilst forty cents, or about 1s. 8d., was the ruling price at the latter. It is of this monopoly and its ruinous consequences that the Canadian so loudly and so bitterly complains. Such indeed is sometimes the want of tonnage in the Canadian seaports, that produce forwarded to tide-water, with a view of being conveyed to Liverpool that season, is not unfrequently detained until the opening of navigation in the following year. The inconvenience of this is great, especially as wheat and flour are perishable commodities, and the exporter loses all the advantages which the English market may in the mean time have offered him.

The remedy for this evil is obviously to throw the

navigation of the St. Lawrence open to the shipping of the world.   This will at once break up the mono-poly which is now so serious a drawback to the trade and agricultural prosperity of the province, at the same time that it will give it every chance of securing to itself that great and growing carrying-trade, to secure which was the chief object of the construction of those expensive works which line the St. Lawrence from Kingston to Montreal.   It was not for the carrying-trade of Canada alone that they were constructed.   If they fail to secure their object the result will be disastrous to the province in a double point of view ; for it will not only lose a great and flourishing trade, which, if fairly dealt with, it has every chance of securing ; but it will also be bur-dened with costly and unproductive works, 'which, instead of being a source of revenue, will turn out to be an annual drain upon the coffers of the province. What say the high protectionists to this prospect ? Will these self-vaunting champions of colonial pro-sperity and greatness maintain a system so ruinous to our finest dependency as this, and all merely to sup-port a stale and tottering theory, and to countenance for a little time longer some antiquated notions as to the only source of England's maritime strength ? Even were the repeal of the navigation laws, in their connexion with the St. Lawrence, a question which was likely to be left entirely to our decision, our true policy, both as regards the mother country and the colony, would be to abrogate them.   But the ques-tion is not one in reference to which we shall be left to consult our own exclusive views and wishes.   The province is bent upon being relieved, and at all hazards, from a restriction which acts so injuriously, not only upon its present fortunes, but also upon its future prospects. And some of those who in Canada are most clamorous for the repeal of the navigation laws, are those whose political sympathies are otherwise most in unison with the views of the protectionists at home. It is not only Liberalism, in Canada—to which the vilest purposes have so frequently but so unjustly been imputed,—which seeks to relieve the St. Lawrence of the restrictions of the navigation laws ; for it is loudly joined in the cry by the humble imitation of imperial Toryism which rears its ridiculous head in the wilds of the province. And when the Canadian asks to be thus relieved, what answer can we now make to him ? We formerly conferred privileges upon him in our markets, which may have compensated, to some extent, for the disadvantages at which in other respects, for the sake of particular interest, we placed him. But these advantages he no longer enjoys. We have deprived him of the price paid him for bearing the burden, and is it fair that he should any longer be called upon to bear it ? He will not consent to bear it much longer, even if we refuse to relieve him of it. And who can blame him for the anxiety which he manifests in reference to the matter, or even the menaces which he is sometimes heard to mutter in connexion with it ? The stake for which he is playing is one of immense magnitude. The trade of the Lakes, for which he wishes to be the great carrier to the ocean, has already attained the value of £30,000,000 sterling : what it will it be in half a century it is impossible to foretell. He has laid himself out at no little cost for the transit trade, and will lose his game if the St. Lawrence below tide-water remains much longer clogged as it is at

present.   By losing it, the expensive works which he
has constructed to secure it will be thrown comparatively unproductive upon his hands, when, instead of
relieving him in whole or in part of the burden of
taxation, as he had every reason to believe they would
do in course of time, they will prove themselves the
cause of additional calls upon his pocket.   Will he,
or should he, submit to this?   Not only justice, but
sound policy also forbids that we should call upon
him to do so; and it is to be hoped, for the sake of
all parties,—for even the shipping interest, if they
bestir themselves properly, have but little to fear
from it,—that the session of 1849 will not pass over
before the St. Lawrence is thrown open to the shipping of the world.*

* The Navigation Laws have since been repealed, and the portion
of the foregoing chapter referring to them, as regards the trade of
the St. Lawrence, is still left in the work in the hope that the information contained in it respecting the two rival routes may be
found both valuable and interesting.   That Canada will immensely
benefit by the change will be found by the experience of this, the
first year of free navigation on her tidal waters.

CHAPTER VI.

FROM BUFFALO TO UTICA, AND THENCE TO MONTREAL BY THE ST. LAWRENCE.

FROM Buffalo I proceeded by steamer, which touched at some of the lake ports of Ohio on the way, to the head of Lake Erie, and up the Detroit River, to the city of Detroit in the State of Michigan. This river is the connecting link between Lake St. Clair and Lake Erie—the former being a small body of water in that neighbourhood, intervening between the latter and the vast volume of Lake Huron, which again is connected with Lake St. Clair by the St. Clair River, both this river and the Detroit being in fact links of the St. Lawrence. The city of Detroit is situated upon the west bank of the river of that name, a little below where it emerges from Lake St. Clair. The Detroit, together with the River and Lake St. Clair, here form the boundary between the State of Michigan and Canada West. From Detroit I proceeded through Canada to the town of London, situated in the midst of a rich agricultural district, the portion of the province lying between Lakes Huron, Erie, and Ontario, with an area about as large as that of England, being as fertile and in every way as desirable a home to the settler as any State in the Union. From London I proceeded by an excellent road, which was planked like a floor for a great part of the way, to Hamilton, at the head of Lake Ontario, whence a sail of fifty miles by steamer conveyed me to Toronto, the capital of the once separate province of Upper Canada. This is a large bustling town, situated on the side of a spacious bay on the northern shore of the lake, and having an extensive commercial intercourse not only with the country behind it, but with all the ports both upon the lake and the St. Lawrence below. It has increased in population as rapidly as any of the American towns to which I have alluded as illustrating the speed with which communities spring up in the New World. Its plan is regular, the main streets running parallel with the shore, and being intersected at right angles by others, which run back from the bay. It is in every respect a pretty town, and its chief thoroughfare, King Street, would be an ornament to any city in the United Kingdom. It is still a species of capital, being the seat of the Court of Chancery and the courts of law for Canada West. Here also is the University, an institution magnificently endowed, but which has hitherto been diverted from its original purpose. It was designed as a provincial institution, but was converted into a sectarian one, the Episcopal church, by a variety of adroit manœuvrings, getting it for a time completely into its hands.   The liberal party in the province are determined to unsectarianize it ; and the liberal ministry now in power at Montreal are devising a measure to be laid before parliament at its approaching session, to place the institution upon a secular basis, when every branch of human learning will be taught in it except theology.   This measure, when introduced, will give rise to considerable excitement, for it is intimately connected with the whole question of the position of the Church in Canada.

But to enter into the particulars of such questions, or to describe minutely what I saw in this part of my excursion, would be alien to the purpose of the present work.   I shall therefore hurry back again to the United States.   There is a daily communication between Toronto and Rochester, and in fourteen hours after leaving the former place I found myself once more on the romantic waters of the Genesee.

After another brief sojourn in Rochester, I proceeded towards the sea-board.   As already noticed, there is a continuous railway from Buffalo to Boston. From Rochester it leads to the village of Canandaigua, which is thirty miles distant.   It was towards sunset when I left, and in about an hour and a half performed the first stage of my journey.   It was the month of August.   The weather was beautiful, and the evening air balmly and delicious.   I remained on one of the platforms of the railway carriage the whole way, enjoying the lovely prospect, through which I was so rapidly driven.   I never witnessed a more gorgeous sunset than that with which the heavens soon glowed behind us. Piles of massive clouds were lounging as it were on the western horizon, their light fleecy fringes glistening as if they had been dipped in silver and gold. Towards the zenith, the sky was of a deep azure; about midway to the horizon it assumed a greenish hue, which became paler and paler as it merged into the brilliant yellow which lay beneath it, which again gave place to the broad belt of flaming vermilion which swept along the horizon, and in which the intervening tree-tops seemed to be bathed. The dazzling picture presented almost every variety of colour and shade, whilst long pencils of white light shot up like bars of sunlight from the horizon to the zenith, spreading like a thin gauze over the brilliant colours underneath, and subduing in some places their intensity. This was the reflected light of the setting sun from the vast body of Lake Erie, which lay directly to the west, and on which the sun was still shining, although it had been for some time below our horizon. These broad bars of light were constantly shifting as clouds drifted in the west over the disc of the invisible sun, the portion of the lake reflecting his lustre at one moment being obscured in shade perhaps the next. It thus appeared as if the scenes were being constantly shifted, which gave to the gorgeous celestial picture a faint terrestrial similitude. There are few places in the world where finer sunsets are seen than in Western New York. In addition to the other causes existing to conjure up the glorious effects, in the midst of which the sun there so frequently descends, are the lakes to the west, lying like huge mirrors, reflecting his lustre to the zenith for some time after he has dipped below the horizon.

I was one of a group of four occupying a portion of the platform, my companions being two Canadians, one quite young, and the other elderly and apparently a retired officer on half-pay, and a Bostonian about thirty years of age, of a more jovial disposition than the great majority of his countrymen. In the course of our conversation, the more elderly of the two Canadians occasionally seasoned his discourse with some of those camp phrases, which do not exactly suit the atmosphere of the drawing-room. On the other side of the platform stood a young man, seemingly under four-and-twenty, with a red pug-nose, grey eye, and altogether a very cod-fish expression, and whose head seemed immovably fixed upon a piece of white cambric, which enclosed a " stiffener " of no ordinary depth. He was evidently a sucking preacher, but to which of the ranting denominations he pertained I could not determine. He did not form one of our group, but listened as attentively to our conversation as if he did. He at length approached us; and addressing himself to our elderly friend, observed that it was quite shocking to hear a man of his years swear so much. The blood mounted to the old man's cheek at this exhibition of impertinence on the part of a perfect stranger and a comparative stripling. For a moment I thought he would have hurled him down the steep embankment which we were then passing; but divining the avocation of his reprover from his greasy white neckcloth, the conventional got the better of the natural man, and instead of striking, he apologised to him, stating that he was sorry that anything offensive should have reached his ears, unwittingly uttered in a conversation not addressed to him. The divine was satisfied, and resumed his former place; but I observed that the Bostonian was almost bursting with suppressed rage. He did not explode until we reached Canandaigua. I was seated in the public room of the hotel, when an altercation suddenly arose in the contiguous lobby. I soon recognised the voice of the Bostonian, who had just caught the parson, and was angrily lecturing him upon his impertinent officiousness.

"What business had you to interfere?" he demanded,—"his conversation was not addressed to you?"

"It is my business to reprove in season and out of season," replied the half-frightened preacher.

"Then by —— you did it out of season that time, I can tell you," said the Bostonian, getting more and more irritated. "He was old enough to be your grandfather," he continued; "besides, you know very well that, if he did swear a little, he didn't mean it."

"That made the matter all the worse," said the preacher.

"All the worse!" repeated the Bostonian, with a choleric laugh; "when I say 'D——n it,' I do mean it; and, according to your doctrine, that is not so bad as to say it without attaching any significance to it."

A loud laugh from the bystanders, who had by this time gathered round, followed this retort, and the discomfited preacher, without uttering another word, entered the public room. The Bostonian followed him to give him a parting admonition, to the effect that he should take care, the next time he reproved, that it was *in* season he did so; by pursuing which course he would do all the less to render himself and his country ridiculous in the eyes of the stranger.

Early next morning I took a stroll through the

village. The small towns, which so profusely dot the surface of Western New York, are in every respect the most charming of their kind in the Union. The country, which is of an undulatory character, abounds with exquisite sites, particularly that portion of it which lies between the Genesee and the upper waters of the Mohawk. The scenery is beautifully diversified by a series of lakes of different sizes, from twelve to thirty and forty miles in length, which follow each other in rapid succession. The land around them is generally well cleared, and the little towns which garnish their banks bespeak a degree of general comfort which is only to be met with in the New World. As you tread their broad and breezy streets, and every now and then catch a glimpse of the elegant white houses with which they are lined, through the waving and rustling foliage in which they are enveloped, you are apt to forget that such a thing as poverty exists, and to give way for the moment to the pleasing illusion that competence is the lot of all. One of the most pleasing features about these towns is their faultless cleanliness. In this respect the Americans are in advance of every other people with whom it has ever been my lot to mingle. An American house, both outside and in, is, generally speaking, a pattern of cleanliness. The American likes to make a good external show, and bestows great care, when circumstances will admit of it, upon the outside of his dwelling. The neat little garden which fronts it is not, as with us, walled from the sight of the public. It is generally bounded towards the street by a low wall, which is surmounted by a light iron or wooden railing, so that the public enjoys the sight of what is within as much as the owner himself. This is what renders not only the rural towns of America, but also the suburbs of its larger cities, so elegant and attractive ; each resident, in consulting his own taste in the decoration of his dwelling, also promoting the enjoyment of the public. How different is the case in our suburbs and country towns ! An Englishman likes to have his enjoyments exclusively to himself; and hence it is that the grounds fronting your " Ivy Cottages," " Grove Villas," and " Chestnut Lodges," are concealed from the passer-by, by lofty, cold, and repulsive walls. There cannot, in this respect, be a greater contrast than that presented by the private streets of an American town, large or small, and those of our own villages and the suburban districts which skirt our great communities. Nor let it be supposed that to this external neatness, in the enjoyment of which the public thus participates equally with the owner, is sacrificed any of the care which should be bestowed upon the management of the residence within. An American is about as domestic in his habits as the Englishman is. His house is, therefore, the private sanctuary of himself and family, and as much attention is generally bestowed upon it, with a view to rendering it comfortable and attractive, as in decorating it externally for the common enjoyment of himself and his fellow-citizens. In point of domestic neatness and cleanliness, the Englishman certainly comes after the American. Would that I could find a high place in the classification for the lower orders of my Scottish fellow-countrymen !

Canandaigua is, in itself, perhaps the most attractive of all the towns of Western New York. There are others with more beautiful sites, but none presenting so fine a succession of almost palatial residences. It is situated on the long gentle slope which descends to the northern extremity of Lake Canandaigua, the most westerly and one of the smallest of the lakes alluded to. The main road between Buffalo and Albany, which passes through it, constitutes its principal street, from every point of which the lake at the foot of it is visible. The street, which is about a mile long, is exceedingly wide, and shaded on either side by an unbroken succession of lofty and magnificent trees. The houses on both sides, which are almost all detached from each other, are some distance back from the street, having gardens in front occupied by grass and flower plots, with clumps of rich green foliage overhead. The finest mansion in the town is the property of a wealthy Scotchman, who has been settled in Canandaigua for upwards of forty years. It is really a superb residence, more like a ducal palace than the dwelling of an humble citizen. The business portion of the town is that nearest the lake, being a continuation of the main and indeed almost the only street of which it boasts.

The country being beautiful and the roads good, I preferred taking the common highway to Auburn, forty miles distant, instead of the railway. I therefore hired a gig, and drove that day to Geneva, sixteen miles from Canandaigua. On leaving the latter, the road led me close to the northern end of the lake, when it suddenly turned to the east, leading over a succession of gentle undulations of the richest country. Before the Erie canal was constructed, and, of course, previous to the introduction of railways, this was the great line of road between the Hudson and Lake Erie. Along it the earliest settlements were consequently made, so that now the aspect

which the country on either side presents is more
like that of an English than an American landscape.
The farm-houses and farming establishments along
the road are large, comfortable, and commodious;
the farmers here being of the wealthier class of practical agriculturists. Some of the houses are built of
brick, others of wood; but whether of brick or wood,
they are all painted equally white, which, in summer
time, gives them a refreshing effect, in contrast with
the clustering foliage which environs them. The afternoon was well advanced when I approached Geneva;
and never shall I forget the beauty of the landscape
which suddenly burst upon my view on gaining the
top of the last hill on the road, about a mile back
from the town. Below me lay Geneva, its white
walls peering through the rich leafy screens which
shaded them. Immediately beyond it was the placid
volume of Lake Seneca, from the opposite shore of
which the county of Seneca receded in a succession of
lovely slopes and terraces. Large tracts of fertile
and well cultivated land were also visible on either
hand; and the whole, lit up as it was by a lustrous
and mellow autumn sun, had a warmth and enchant-
ment about it such as I had but seldom beheld in
connexion with a landscape.

Geneva is a much larger town than Canandaigua;
and I know no town in America or elsewhere, with
so charming a site. Lake Seneca, like all the other
lakes in this portion of the State except Oneida, is
long and narrow, and lies in a northerly and southerly
direction. On its west bank, at its extreme northern
end, stands Geneva. The business part of the town
is almost on the level of the lake; the bank, which
is clayey, high and abrupt, suddenly dropping at the

point where it is built. It is on the high bank, before it thus drops, that the remainder of the town is built, most of the houses of which command a view of the lake. The most eligible residences are those which skirt the lake, with nothing but the width of the road between them and the margin of the bank. They have an eastern aspect, and nothing can exceed the beauty of the view commanded from their windows, as the morning sun rises over the landscape before them.

I was so delighted with Geneva that I prolonged my stay there for two days longer than I had at first intended. On the evening of my arrival I took a small boat and went out upon the lake. It is about forty miles long, but scarcely a mile wide opposite Geneva. The air was still, but the western sky looked angry and lurid. As it gradually blackened, a fitful light every now and then faintly illuminated the dark bosoms of the massive clouds, which had now made themselves visible in that direction. As they stole higher and higher up the clear blue heavens, the illumination became more frequent and more brilliant, and nothing was now wanting but the muttering of the thunder to complete the usual indications of a coming storm. I was then some distance up the lake, and made as speedily for town as possible. When I reached it, innumerable lights were gleaming from its windows upon the yet placid lake, whose dark, still surface was occasionally lit up for miles by the lightning which now coruscated vividly above it. The first growl of the distant thunder broke upon my ear as I stepped ashore; and, pleased with my escape, I hurried, without loss of time, to the hotel. In a few minutes afterwards the progressing storm burst over the town, and the dusty streets soon ran with

torrents of water. The effect upon the lake was
magnificent.   It was only visible when the lightning,
which now fell fast on all sides, accompanied by
awful crashes of thunder, gleamed upon its surface,
and seemed to plunge, flash after flash, into its now
agitated bosom.   You could not only thus distinguish
the dark leaden waters, with their foaming white
crests, but the shore on the opposite side for a considerable distance inland, and on either hand.   The
whole would be brilliantly lighted up for a moment
or two, after which it would relapse into darkness, to
be rendered visible again by the next succession of
flashes which fell from the black and overcharged
heavens.   In half an hour it was all over, when the
scene displayed itself in a new aspect, veiled in the
pale lustre of the moon.

A steamboat communication is daily maintained
during summer between Geneva and the southern
end of the lake.   On the following day I sailed
about half way up, and rode back to Geneva in the
afternoon by the bank.   Both shores, which were at
some points low and flat, and at others elevated and
rolling, were highly cultivated, which is indeed the
case with the whole of this section of the State
almost from Lake Ontario, south to the Pennsylvania
line.   I left Geneva after a sojourn of three days,
and with recollections of it which will never be
effaced.

Passing around the head of the lake I crossed the picturesque and rich agricultural county of Seneca, lying
between Lakes Seneca and Cayuga, which are about the
same size, and stretch in long parallel lines in the same
direction. After a drive of about three hours' duration I
found myself descending upon Lake Cayuga, at a point a

few miles from its northern extremity. I had scarcely begun to puzzle myself as to how I was to get across, when the means of passing the lake was gradually presented to my astonished vision. A bridge of nearly a mile in length spanned its volume at this point, the opposite end of which first came in view; nor was it until I had reached the lake, that the whole length of this stupendous viaduct was visible to me. It was constructed of wood, and laid upon a series of wooden piers, which lifted their heads in long succession but a few feet above the level of the water. It was in every way a more singular construction, in my estimation, than the long bridge over the Potomac at Washington. There was a similar structure a little to the left, over which the railway passed; and before I had half crossed that which was in the line of the common highway, the eastern train shot from an excavation in the opposite bank, and went panting over the railway bridge at unabated speed. Lake Cayuga is the dividing line between Eastern and Western New York, and in little more than an hour after crossing it, I found myself in the lovely town of Auburn.

I stayed here for the night, and visited the State prison, in other words, one of the State penitentiaries. But so much has already been written about the prisons and prison discipline of the United States, in which the penal establishment at Auburn has invariably been described, that I need not here trouble the reader with an account of it. It is, of course, surrounded by high walls, and is in its exterior both neat and elegant, looking half like a prison and half like a palace.

Next morning I betook myself once more to the

railway on my way to Utica. Our first stage was Syracuse, the capital of the county of Onondaga, one of the most populous as well as one of the finest agricultural counties in the State. On approaching Syracuse, which is an open, airy, handsome town, divided into two sections by the Erie canal, which runs through it, we passed the great salt works of Salina. The salt springs of the district appear to be inexhaustible. They are the property of the State, which derives a good annual revenue from leasing them. Enormous quantities of the finest salt are yearly made here, both for home consumption and for exportation. There are some purposes, however, such as curing, for which it is not available, and for which it comes but partially in competition with the rock salt at Liverpool.

From Syracuse to Utica the distance is fifty miles. Rome lies in the way. Some little time after we had performed half the journey, the railway led for nearly five miles in one continuous straight line, through a dense forest, which kept in perpetual shade a large tract of low marshy soil. At the extreme end of the long vista which thus opened through the wood, I could discern a white steeple rising over a circumjacent mass of bright red brickwork.

" What place are we now approaching," I demanded of a fellow-traveller.

" Rome," said he. " I live to Rome myself; it's gettin' to be quite a place." I thought it was high time that Rome did so.

We were now in the valley of the Mohawk, which stretches eastward to the Hudson; and in less than an hour after leaving Rome I found myself in Utica, the capital of Central New York.

The reader will be astonished at finding so many places in this modern scene named after those with which his schoolboy reminiscences are so intimately associated. They are jumbled together in ludicrous juxtaposition; sometimes one and the same county in the New World containing two towns, living in peaceable intercourse with each other, for which there was scarcely room enough on two continents in the Old. New York, in particular, abounds in places having classical appellations; a rather singular circumstance when we consider the many beautiful and expressive Indian words which it might have appropriated to the purposes of a civic nomenclature. Proceeding eastward from the Falls, one of the first places you meet with is Attica, from which a single stage brings you to Batavia. A little to the east of Rochester you pass through Egypt to Palmyra, whence you proceed to Vienna, and shortly afterwards arrive at Geneva. Ithaca is some distance off to the right, whilst Syracuse, Rome, and Utica follow in succession to the eastward. It is a pity that the people in the New World should not content themselves with indigenous names. They are quite as pretty, and would in many cases be more convenient than those which have been imported. The inconvenience arises not so much from naming places after cities which have passed away, as after those who are still extant and flourishing. There is a New London on the Thames in Connecticut, and there is a London on the Thames in Western Canada. There is scarcely a town of any note in Europe but has scores of namesakes in America, whilst the Indian dialects are replete with significant and sonorous terms. What a happy change did "Little York" make when it called itself Toronto!

Utica is a fine town, with from twelve to fifteen thousand inhabitants.   Its importance in a commercial point of view has greatly declined since the construction of the railway connecting the Hudson with the West.   Previously it occupied the position of a kind of advanced post of New York, from which the interior was chiefly supplied during the winter.   The communication, however, being now so rapid and direct with the sea-board, its business is chiefly of a local character.   The Erie canal passes through the centre of it.   It is crossed at right angles by the broad and noble Genesee-street, the *coup-d'œil* of which, as seen from the canal bridge, is exceedingly striking.

When in Utica a few years previously, I strolled into the supreme court of the State, which was then in session.   Neither the justices on the bench nor the members at the bar wore any particular dress to distinguish them from the spectators.   When I entered, a venerable looking man, with thin grey locks, a high forehead, and altogether an engaging countenance, was addressing the court, arguing a demurrer.   The case was that of Cooper (the novelist) *v.* Stone (the editor of a New York paper), the action having been brought for a libel, published by the latter, in reviewing the former's " Naval History of America."   It appeared that the defendant had demurred to the declaration filed in the case, and the advocate was now engaged in maintaining its sufficiency in law.   There was a good deal in his appearance and manner which induced me to think that he was not one of the fraternity in the midst of whom he was then placed, whilst the frequency with which he moistened his parched palate with the orange which lay on the table before him, indicated that he was " unaccustomed to public speaking." The case at the time excited great interest, and I remained for nearly an hour listening to the argument. It appeared that the defendant, Stone, had himself many years ago published something replete, as the advocate contended, with blunders, which an over zealous critic might have turned to some account. Having adduced two or three of these, which sufficed for his purpose, he insisted that "those who lived in glass houses should not throw *stones*." It was not until I had left the court that I was given to understand that the advocate in question was the great American novelist himself. His appearance in a forensic capacity, in thus pleading his own cause, did him considerable credit.

The tourist should always make a halt at Utica, that he may visit the Falls of Trenton in its neighbourhood. They are fourteen miles to the north of the city, and are approached in summer by a road which is tolerably good. On the morning after my arrival in Utica, I hired a conveyance, and proceeded to them. Immediately on leaving the city, which is built upon its right bank, I crossed the Mohawk, here a sluggish stream of very insignificant dimensions. Moore must have seen it much lower down, ere he could speak of the "mighty Mohawk." The road then led, for nearly a couple of miles, over a tract of rich bottom land, as flat as the fertile levels of the Genesee valley. It then rose, with but little intermission, for the next six miles, by a succession of gentle slopes, which constitute the northern side of the valley of the Mohawk. On reaching the summit, I turned to look at the prospect behind me. It was magnificent. The valley in its entire breadth from bank to bank lay beneath me; whilst an extensive range of it in an easterly and westerly direction came also within the scope of my vision. As far as the eye could reach it was cultivated like a garden, whilst far beneath, on its lowest level, on the opposite bank of the river, the serpentine course of which I could trace for miles, lay Utica, its skylights and tin roofs glistening like silver in the mid-day sun. The opposite slope of the valley was dotted with villages, some of which were plainly visible to me, although from twelve to twenty miles distant in a straight aërial line.

To the north the view was also extensive, but of a more sombre cast. The country was less cleared, and plain after plain seemed to stretch before me, covered by the dark gloomy pine. For the rest of the way to Trenton the road descended by a series of sloping terraces, similar to those by which it had risen from the valley.

After taking some refreshment at the hotel, which is beautifully situated, spacious, and comfortable, and which at the time was full of visitors, I descended the precipitous bank to look at the Falls. I dropped by a steep zigzag staircase, of prodigious length, to the margin of the stream, which flowed in a volume as black as ink over its grey rocky bed. Frowning precipices rose for some distance on either side, overhung with masses of rich dark green foliage. A projecting mass of rock, immediately on my left, seemed to interpose an effectual barrier to my progress up the stream. But, on examining it more carefully, I found it begirt by a narrow ledge overhanging the water, along which a person with a tolerably cool head could manage to proceed by laying hold of the chain, which either public or private beneficence had fastened for his use to the precipice on his left. On doubling this point, the adventurous tourist is recompensed for all the risks incurred by the sight which he obtains of the lower fall. It is exceedingly grand, but not on the same scale of magnitude as the Falls of the Genesee. It is the accompanying scenery more than the cataract itself that excites your admiration. The opposite bank is high and steep, but not precipitous, and is buried in verdure; whilst that on which you stand rises for about 200 feet like a grey wall beside you. The fall occupies an angle here, formed by the river in its course. In turning it, you take the outer circle, climbing from ledge to ledge, the friendly chain again aiding you every now and then in your course, until you find yourself on a line with the upper level of the fall. Here the cataract next in order comes in full view; and a magnificent object it is, as its broken and irregular aspect rivets your attention. It is by far the largest fall of the whole series, being, in fact, more like two falls close together than one. There are two successive plunges, the first being perpendicular, and the second a short but fierce rapid foaming between them, being divided into a succession of short leaps by the jagged and irregular ledge over which it is taken. By the time you attain the level of the top of this fall, by climbing the still steep and slippery rock, you reach the wooded part of the bank. Your progress is now comparatively easy, the path occasionally leading you beneath the refreshing shade of the large and lofty trees. Below, you had the native rock rising in one unbroken volume precipitously overhead; but you have now on either side what may be regarded more as the ruin of rock, the trees with which both banks are covered springing, for the most part, from between huge detached masses, which seem to have been confusedly hurled from some neighbouring height. The channel of the stream is broad and shallow up to the next fall, which, in its dimensions and appearance, resembles a mill-dam. Above, the river contracts again, until in some places it is only a few yards wide, where it foams and roars as it rushes in delirious whirl over its rocky bed. A little way up is the last cataract, the most interesting in some respects, although the smallest of all. To pass it you have to turn a projecting point, the narrow footpath around which brings you almost in contact with the rushing tide, as it bounds over the ledge. Here the chain is almost indispensable for safety. A melancholy interest attaches to this fall, from some sad and fatal accidents which occurred at it before the chain was placed where it is. The gorge through which the West Canada Creek, such being the name of the stream, here forces its way, is about two miles in length. Between the upper fall and the small village at the northern extremity of the chasm, the channel of the river is a succession of rapids and dark eddying pools covered with patches of dirty white foam. I managed with great difficulty, and with the aid of a guide, to ascend it, returning to the hotel by the open road leading along the top of the bank.

Trenton is well worth a visit, and the tourist may enjoy himself for some days in its neighbourhood. There is much about it to remind one of the wild and romantic scenery of Campsie Glen, in the neighbourhood of Glasgow. There is the same succession of falls and rapids, but the stream is larger, and the cataracts on a greater scale, at Trenton than at Campsie.

Instead of returning to Utica, and descending the valley of the Mohawk by railway to Albany, I determined to strike across the country from Trenton to the St. Lawrence, descend that river to Montreal, and find my way thence, by Lake Champlain, into the United States. After a rather tedious day's journey I reached Watertown, and soon afterwards found myself on the American bank of the St. Lawrence, just as it emerges from Lake Ontario. I crossed it at once into Canada, landing in the town of Kingston. For the reason stated in briefly adverting to Toronto, I shall make no particular allusion to that portion of the remainder of my journey which led me through Canada.

There were two sets of steamers plying between Kingston and Montreal; one descending all the rapids, and the other stopping short at those of Lachine, close to the latter city. I selected one of the former, determined, when I was at it, to shoot them all.

Rounding the point in front of the town, on the top of which stands the impregnable little fortress of Fort Henry, we shaped our course down the river, the surface of which was broken into several narrow and winding channels, by the many islands with which it was studded. The extraordinary group called the "Thousand Islands" commences about twelve miles below Kingston, and extends to Brockville, about sixty miles further down. As we approached, the St. Lawrence seemed to be absorbed by the land before us, which was covered with wood, and appeared to sweep in an unbroken line across the channel of the river. It was not until we were close upon it that the whole mass seemed suddenly to

move, and the change effected was as complete as
when the scenes are shifted in a theatre. Some
portions appeared to be drawn aside in one direction,
and others in another, until that which, a few minutes
before, looked like one solid mass of earth, seemed as
it were to be suddenly broken into fragments, divided
from each other by innumerable channels, varying as
much in their dimensions as in the directions which
they took. We made for the widest, and plunged
into the labyrinth. We sometimes gave the islands
a good berth, but at others had them so close on
either side of us that I could almost jump ashore from
the paddle-boxes. In naming them the " Thousand
Islands," people came far within the mark as regards
their actual number, which, I understand, exceeds
fifteen hundred. They are of all sizes, from an area
of 600 acres to that of an ordinary dining-table. They
rise but a few feet above the surface of the water,
some being covered with large timber, and others
with stunted shrubbery, whilst some, the smallest of
the group, have no covering whatever, nothing but
the bare rock peering above the river. For the whole
way through them you are subjected to a series of
surprises. You puzzle yourself every now and then
as to where your next channel is to be, when perhaps
the advance of a few yards more solves the difficulty,
one, two, or more, suddenly opening up on the right or
on the left, as the case may be. This singular group
seems to be the remains of a low ridge of earth and
rock which lay in the river's course just as it emerged
from the lake, the accumulated waters of which at
length burst through the impediment without carrying it bodily off, but making for themselves sufficient
room to escape.

From Brockville, for some distance down, the broad channel of the river is free from impediment. Twelve miles below is the town of Prescott, and on the American bank opposite that of Ogdensburg. During the insurrection which broke out in Canada in 1837, a piratical expedition landed from the latter place at Windmill Point, a few miles below Prescott. They were under the command of a Pole of the name of Von Shultz. I am not aware whether he was ever a recipient of British bounty or not, but perhaps Lord Dudley Stuart could tell. The bucaniers were defeated, and Von Shultz was hanged.

On leaving Prescott we crossed to Ogdensburg for passengers. This town is built at the mouth of the Oswegatchie, a rapid stream, with so dark a current that, on entering the St. Lawrence, it seems to run with ready-made porter. Gliding down the river we were soon in the midst of islands again, and found ourselves ere long at the commencement of the rapids.

The first two or three which we passed were not sufficiently formidable to cause more than a slight ripple on the surface; but by-and-by we approached the great rapid, that called the Long Sault, and preparation was made for its descent. Even those accustomed to shoot it seemed to grow more and more excited as we approached; it was no wonder, then, that a novice like myself should partake largely of the feeling. We touched for a few minutes at Dickenson's landing, a little above the rapid, and already alluded to as being at the upper extremity of the St. Lawrence canal, constructed to enable vessels not built for descending the rapid, to avoid it. On getting afloat again the ladies retired to the cabin, half-frightened at what was before them, and determined, at least, not to witness the danger. I took my post upon deck, where I resolved to remain until the exciting episode was over. The rapid was in sight. Independently of the fact that I was about to shoot it, it was an object of the highest interest to me, for who has not heard of the rapids of the St. Lawrence? In my mind they were associated with my earliest reading reminiscences. We were close to the Canada shore, some wooded islands intervening between us and the American bank. The rapid commenced amongst the islands, but did not exhibit itself in its full force and grandeur until it emerged from them into the clear and somewhat contracted channel immediately below. Throughout its whole length it is much more formidable on the Canada than on the American side. It was by the latter alone, previously to the completion of the canal, that the barges which were used in the navigation of the river could ascend, on their way from Montreal to Prescott. It sometimes required fourteen yoke of oxen to tow an empty barge slowly against the current, not where it was most impetuous, but close to the shore, where its force was comparatively small. It was by the Canada side that we were to descend the rapid, which leapt, foamed, and tossed itself wildly about, immediately in front of us. As far as we could see down the river, the dark leaden-looking water was broken into billowy masses crested with spray, like the breakers upon a low rocky shore, stretching far out to sea ; whilst the roar with which the delirious current was accompanied, was like the sound of a cataract hard by. For nearly a quarter of a mile above the rapid the current ran smoothly, but with great velocity, which increased as it approached the line at which the channel dipped still more, agitating the mighty volume, which seemed to tear itself to pieces against the sunken rocks, over which it dashed with impetuous speed. A period, as it were, of breathless expectation ensued, from the time of our entering upon the preliminary current, until we crossed the line in question. The steamer seemed here to take its race for the plunge which it made from the smooth into the broken current. To one unaccustomed to such a scene, a moment or two of semi-stupefaction ensues, after getting fairly within the embraces of the rapid. It seemed to me at first that we had suddenly been brought to a halt, and were standing still, with the water boiling and surging around us in a mighty caldron, whilst islands, mainland, rocks, trees, houses, and every fixed thing ashore seemed suddenly to have been loosened from their foundations, and to be reeling around me. On becoming more collected I discerned the real state of things; the steamer was shooting like an arrow along the stormy descent, lashing the angry waters with her lusty paddle-wheels to give her steerage-way. She thus rushed on for miles in the course of a few minutes, the objects ashore flitting by us as do those which line a railway. By-and-by we reached a point where the current, although yet greatly agitated, was comparatively tranquil, when the very steamer seemed to breathe more freely after her perilous race. On looking around me, the islands were gone, and the broad and broken channel was no longer to be seen, the banks had fallen from their well-wooded elevations almost to the water's edge, the stream was contracted—it was placid in front of us, but wildly agitated behind—in short the whole scene had changed. The whole looked like a troubled dream, and it was some time ere I could recall, in their proper succession, the different incidents which marked it.

We soon afterwards turned a point which shut the rapid from our view, and in about half an hour more were at Cornwall, the frontier town of Canada West. Here we stayed for the night, resuming our journey at a very early hour in the morning.

Shortly after leaving Cornwall, we emerged from a mazy but very interesting channel upon the broad and placid volume of Lake St. Francis. It is about fifty miles in length, and is studded at its upper end with a pretty cluster of islands. Indeed, its whole surface is more or less dotted with islands, on one of which the highlanders of Glengary, a county in Canada West, bordering the lake on the north, have erected a rude conical monument of unhewn stone, in honour of Sir John Colborne, now Lord Seaton, who took such prompt and decisive measures for the suppression of the insurrection in 1837 and 1838. The boundary line between the two provinces is but a little way below, and this monument has been rather ungraciously raised in sight of the *habitans*. Passing the boundary, we found ourselves at once amongst the French settlements; the long, low, wooded bank, which loomed upon us in hazy outline to the south, being a portion of the fertile seigniory of Beauharnois——both banks of the St. Lawrence being now within British jurisdiction. At the foot of the lake, on its northern shore, stands the little French Canadian town of *Coteau du Lac*. And here commences the series of rapids, occurring, with but little intermission, between Lake St. Francis and

Lake St. Louis. The channel of the river was once more impeded with islands, a screen of which stretched across it on its emerging from the lake, which almost entirely concealed the rapids below from the town. The first rapid which occurred was that known as the Coteau; it was short, but exceedingly impetuous, and we seemed to clear it almost at a bound. Between that and the Cedars, the next rapid, the current was deep and strong, but the surface was unbroken. We shot at a very swift pace by the village of the Cedars, situated at the head of the rapid on a pretty bend of the high bank on our left. On turning a point immediately below it, the rapid became visible, and in ten minutes more we were at the foot of it. The channel is here very wide again, and the rapid pours between the islands which stud it, as well as between them and the banks. Here the strongest point of the rapid is on the southern side, where the water, as it leapt from rock to rock, rose high in the air, crested with white foam which looked like snow wreaths in the distance. For a mile or two the channel was again unimpeded and smooth, though the current was still very strong. We then approached another group of islands, with seemingly several channels between them. Here I had proof, in the tossing and angry waters before me, that we were about to shoot another rapid, known as the Cascades. We were soon in the midst of it, when we seemed to glide down a hill into the tranquil volume of Lake St. Louis. On looking to the left, as we pursued our way down the lake, I perceived a noble estuary stretching for miles in a north-westerly direction, with a line of blue hills faintly traced along the sky beyond it. This was the mouth of the Ottawa, which in uniting here with the St. Lawrence forms Lake St. Louis. I looked with much interest upon the spot where two such mighty streams peacefully mingled their confluent waters; one issuing from the distant Lake Superior, and the other rising in the remote territories of the Hudson's Bay Company. The St. Lawrence came foaming and roaring down its rugged channel into the lake, whereas the Ottawa stole gently into it, in a broad and scarcely perceptible current. Some distance above, however, its channel is interrupted, like that of the St. Lawrence, by a succession of rapids.

A run of two hours sufficed to bring us to the lower end of Lake St. Louis, and here we prepared to descend the rapid of Lachine, the last great rapid of the St. Lawrence. It is by no means so formidable in point of size as the Long Sault Rapid, but it is more perilous to navigate. The larger steamers which descend the river do not attempt it, stopping short at the village of Lachine, whence passengers are conveyed for nine miles by coach to Montreal. To almost every one on board was assigned his place ere we reached the top of the rapid. I was near the pilot, who was an old Indian, and I could not but mark the anxiety with which he watched the progress of the boat as he managed the wheel, seeming desirous of getting her into a particular line ere she commenced her headlong race. As she shot down she rocked heavily from side to side, and when about half-way to the foot of the rapid, her keel grated the rock at the bottom. The concussion was severe, and as she lurched over a little before getting again fairly afloat, the glancing waters leapt upon deck, soaking many of the passengers to the knees. There were few on board who were not frightened, and some, the ladies particularly, screamed outright. The alarm was but momentary, and before we had recovered our equanimity we had emerged from the islands which the rapids here also encircle, and had the noble capital of Canada full in view on our left, nestling beneath the hill which gives it its name, close to the north bank of this the main branch of the stream.*

* Montreal is no longer the capital of the Province. The outrageous conduct of the factious cliques, who opposed the Rebellion Losses Bill, induced the Governor General to remove the seat of Government to Toronto. It appears that it has since been decided to divide the honour between Toronto and Quebec. The policy of thus having a peripatetic Government, oscillating between two points about 600 miles apart, is very questionable. Toronto is, in every way, better fitted than any other town, for being the Capital of the United Provinces.

CHAPTER VII.

FROM MONTREAL TO SARATOGA, ALBANY, AND WEST POINT.—MILITARY SPIRIT AND MILITARY ESTABLISHMENTS OF THE UNITED STATES.

Departure from Montreal.—The "Tail of the Rapid."—Appearance of Montreal from the river. — Laprairie. — Lake Champlain.— Rouse's Point.—Banks of the Lake.—The Representative System in Vermont.—The "Devil's Elbow."—Whitehall.—Saratoga.— Life at the Springs.—The Table-d'hôte.—Troy.—Albany.—Accident on the Hudson.—Stay at Hyde Park.—Scenery of the Hudson.—The Highlands.—West Point.—The Military Academy.— The Military Spirit in America.—Exaggerated notions respecting it.—Military Establishments.—Cost of the Army and Navy.— Contrast presented by the Military Expenditure of the United Kingdom and the United States.—The Defence of our Colonies.— Their Mismanagement.—Our Military Force in Canada.

I STAYED for several days at Montreal, to which I returned after having descended the river for nearly 200 miles more to Quebec. It was a beautiful morning, near the close of August, when I finally bade adieu to the former city on my way back to the United States. To take the railway to St. John's, I had to cross the ferry to Laprairie. In crossing, the ferry-boat skirted what is called the tail of the rapid, and was tossed considerably about by the yet uneasy river. I looked with some interest at the rapid itself, raging above us, down which I had been hurried but a few days before, and in shooting which I must confess that I was somewhat disconcerted, although not without cause.

As viewed from the river, the position of Montreal is exceedingly commanding. It is a large town, with a population of about 60,000, being chiefly built of stone, and lying close to the margin of the river. A line of solid stone quays fronts the city, which rises in a dense mass behind them, the background of the picture being filled up by the hill behind, which is skirted with orchards and dotted with villas. One of the most prominent objects in the outline of the town as seen from the river is the Catholic Cathedral—with the exception of that in Mexico, the finest ecclesiastical edifice on the continent. There is a new and an old town, much of the former reminding one of some portions of Havre or Boulogne. The new town is more symmetrically laid out, the streets being broad, and the architecture of a superior description.

I took the railway at Laprairie, and after a short ride over a tolerably well cultivated country, found myself at St. John's, not far from the northern extremity of Lake Champlain. I was desirous of visiting this lake, on account of the important part which it played, not only in our two wars with America, but in the conflicts which so frequently took place between the French and English colonies, whilst England still ruled from the Kenebec to the Savannah, and whilst Quebec and Montreal were French towns, and "New France" comprised within its vast limits fully two-thirds of a continent, on which France has no longer a footing.

I went on board the American steamer, which was waiting for us at St. John's, and in a few minutes afterwards we were under weigh. The steamer was, in her appearance and all her appointments, one of the most magnificent of the kind I had seen in America. She was a floating palace.

Shortly after getting upon the lake, which was very narrow at the top, as it is indeed throughout its entire length, which is about 150 miles, we crossed the 45th parallel of latitude, here forming the boundary line between Canada and the United States. Rouse's Point, which has figured somewhat of late in the diplomacy of the two countries, was pointed out to me as we approached the boundary. I must confess that it appeared to me that nothing but an overpowering necessity could justify the cession of so important a military position. The navigable channel of the lake is here not many yards wide, and is so situated as regards the point which projects for some distance into it, that a vessel approaching it would be for a considerable time exposed to a raking fire from stem to stern, before she could come abreast of it and make use of her broadside. The same would be the case were she to approach it from the other side; so that a ship proceeding from Canada to the United States, and *vice versâ*, would not only be thus exposed on approaching, but also on receding from the fort. For some time after the boundary line was originally agreed upon, it was supposed that the point fell within the American side of it; and, acting on this supposition, the American government proceeded to fortify it. It was found, however, on accurate observations being taken, that it fell upon the Canada side of the line, and the works in progress were of course immediately discontinued. It remained without anything further being done to it in the hands of the British government, until by the treaty of Washington, negotiated on our behalf by Lord Ashburton, it was ceded in perpetuity to the United States. And what was the equivalent? The smaller moiety of the debateable land in Maine, all of which we claimed to be our own. Better have ended the controversy by giving up the whole of the disputed territory, and retaining Rouse's Point, which had nothing to do with it, and respecting which there was no controversy. We should have been quite as well off, even on the route between Halifax and Quebec, had we ceded to the Americans the line for which they contended, as we are by the retention of only that portion of the disputed ground which has been left us. The territory in dispute was valuable to us only inasmuch as its possession would have enabled us to construct a pretty direct military road, railway, or common road, between Montreal, or Quebec, and Halifax. It was valuable to the Americans only inasmuch as their possession of it would prevent us from constructing such a road. The portion of it which has been ceded to us does not strengthen us, whilst their parting with it does not weaken them. The road, to be of any value as a military road, must not only lie entirely within our own territory, but should be so situated as to be commanded by us at all points. But, should it ever be now built, it will be commanded for nearly half its length by the Americans. A road leading through that portion of the country must be virtually, in time of war, in possession of the party who can keep the largest force most easily near it in the field. Projecting, as the northern part of Maine now does, like a wedge between the two extremities of this intended road, to keep it within our own territory it would be necessary to carry it, by a very circuitous route, up almost to the St. Lawrence, so as to turn the portion of the State mentioned, which approaches so near the banks of that river. A British force stationed

along the line of the road would be at every disadvantage for want of a good basis of operations; whereas
an American force would have the sources of its
supply immediately behind it, and a secure means of
retreat, should retreat become necessary.   In addition
to this, as regards the road, the latter would have the
advantage of acting, as it were, from the centre of the
circle, an arc of which the road would describe around
it.   The Americans could thus command it at any
point for hundreds of miles, and possession of it at
one point would be tantamount to a possession of the
whole line.   Yet it is for this that we have given up
the key of Canada in the neighbourhood of Montreal
—a point commanding the most direct and practicable
highway between Montreal and Albany, between the
St. Lawrence and the Hudson.   It may be urged that
it matters but little, as we are not likely to go to war
again with the United States.   I trust we are not,
but stranger events have happened within the year
than the breaking out of such a war.   Besides, how
much longer we are to retain Canada as a dependency
is a question.   Should she merge into the Union, the
value of Rouse's Point as a military position will
materially decline; but should she take an independent position on the continent, as this country
would naturally wish her to do in case of separation,
she would feel herself much disabled, in a military
point of view, by the loss of the point in question.
In time of war it would throw open to the enemy the
highway to her capital.   It was our duty, in consulting our own convenience, to have some regard
for the interests of our dependency; and the time
may yet come when Canada will have reason bitterly to
regret the ill-starred liberality of the mother country.

Having passed Rouse's Point, we made down the lake at a very rapid rate. The sail was exceedingly interesting, the lake being narrow, and the banks varying a good deal in their outline, being in some places deeply wooded, and in others highly cultivated. With the exception of the small portion at its northern end, lying within the Canada line, Lake Champlain is entirely within American territory, forming, for about 150 miles, the boundary between New York and Vermont—the former constituting its west, and the latter its east bank. The New York shore is generally much lower than that of Vermont, which, as you approach the middle of the lake, swells into bold and sweeping undulations from its very margin. Burlington, the chief town of Vermont, occupies a fine sloping site on the east bank, about half way down the lake. From this point, some idea may be formed of the nature of the " Green Mountains " which traverse this State, and from which it takes its name. They are generally covered with masses of pine forest, which impart to them that dark-green sombre hue, which, in this respect at least, would render them fit associates for the uplands of Montenegro.

Vermont, though a strong Whig State in its politics, is one of the most democratic of all the States in its polity. It tried for a time the experiment of a single chamber; and there cannot be a better argument in favour of a double one than the fact, that this quiet, orderly, thrifty, decorous, and sober State soon found it advisable to resort to it. Whether it is the case now or not, I am not able to say; but the time was when each township in the State was represented in the Lower House. It so happened, that in one township there were then but three inhabitants, a father, his son, and a farm servant. To avoid the excitements consequent upon an election, they soon came to an arrangement to go time about to the legislature : the father first, to maintain the interests of property in possession; the son next, to see that expectancies were duly cared for ; and then the servant, to vindicate the rights of labour. How long the arrangement lasted I was not informed.

After leaving Burlington, night rapidly closed around us. Early next morning we came to a halt, and shortly afterwards a great hubbub arose upon deck, as if we were about suddenly to be run down by something. I dressed hastily, resolved to die standing, if at all necessary ; but on getting above I found that all the uproar had been caused by the operation of turning a point near the southern end of the lake, called the "Devil's Elbow." We here entered by a very narrow strait a small branch of the lake, at the top of which was Whitehall, our destination ; and as the boat was very long, and the turn very sharp, it was necessary to pull her round by means of ropes. I asked a fellow-passenger why the point which we were thus awkwardly doubling was called the Devil's Elbow more than anybody else's; but he remarked that he could not tell, unless it were that it took "such a devil of a work to get round it." Whitehall is most romantically situated within a few rifle shots of this piece of Satanic anatomy. On landing we proceeded to the hotel by omnibus, our driver being a young man, dressed in superfine black, and wearing a swallow-tailed coat. So far as dress went, he might have stepped from his box into the ball-room.

From Whitehall I proceeded by stage to Saratoga, on my way to Albany, the distance being about seventy miles. With the exception of the land here intervening between Lake Champlain and tide-water on the Hudson, the narrow strip between the lake and the St. Lawrence is the only impediment which exists to a complete water communication between New York and Montreal. It is this that renders Rouse's Point so valuable a point of offence and defence. Between Whitehall and Saratoga the country was rolling and elevated, and very generally cultivated. In some parts a deep heavy clay was on the surface, in others the soil was light and rather sandy. We proceeded by Glen's Falls on the Hudson, crossing the river just below them. The town to which they have given their name is neat and pretty, and the whole neighbourhood is charming, but the Falls themselves are nothing as compared to those of Trenton, or of Portage on the Genesee. It was evening when we arrived at Saratoga; and glad was I to alight at the hotel, after a hot dusty ride inside a closely-packed coach.

Saratoga has lately been losing caste, but it is still, to a considerable extent, a place of fashionable resort. For a time the " select " had it all to themselves, but by-and-by " everybody " began to resort to it, and on " everybody " making his appearance the " select " began to drop off, and what was once very genteel is now running the risk of becoming exceedingly vulgar. The waters are held in considerable repute as medicinal; but of the vast crowds who flock annually to Saratoga, but a small proportion are invalids. The town is very elegant, the main street being enormously wide, and shaded by trees. The hotels are on a very great scale, and so are their charges. At this, however, one cannot repine, seeing that it is everybody's business to make hay when the sun shines. It scarcely shines for three months for the hotel-keepers of Saratoga, the crowds of flying visitors going as rapidly as they come with the season. For nine months of the year Saratoga is dull to a degree—duller, if possible, even than Washington during the recess of Congress. Suddenly the doors are opened—the shutters are flung back from the windows—curling wreaths of smoke rise from the long smokeless chimneys—and the hotels seem suddenly to break the spell that bound them to a protracted torpidity. A day or two afterwards, a few visitors arrive, like the first summer birds. But long ere this, from the most distant parts of the Union people have been in motion for " the Springs," and scarcely a week elapses ere the long-deserted town is full of bustle and animation, and ringing with gaiety. A better spot can scarcely be selected for witnessing the different races and castes which constitute the heterogeneous population of the Union, and the different styles of beauty which its different latitudes produce. I stayed several days and enjoyed myself exceedingly, and seldom have I seen together so many beautiful faces and light graceful forms as I have witnessed on an August afternoon upon the broad and lengthy colonnade of the principal hotel.

I was so fortunate as to meet at Saratoga with a Canadian friend, who had been my fellow-voyager across the Atlantic. The gaiety of the place is infectious, and we soon entered into it with the same eagerness as those around us. Saratoga society is not encumbered with conventionalities. To society around it, in its general acceptation, it is what the undress boxes are to the more formal circle beneath. You make acquaintances there whom you do not necessarily know, or who do not know you elsewhere. The huge pile constituting the hotel covered three sides of a large quadrangle, the fourth side being formed by a high wall. The whole enclosed a fine green, on a portion of which bowls could very well be played. The three sides occupied by the building were shaded by a colonnade, to protect the guests from the hot sun. This part of the establishment was generally appropriated by them, where they lounged on benches and rocking chairs, and smoked and drank both before and after dinner. The meal just mentioned was the "grand climacteric" in the events of each day. A few families who visit Saratoga dine in their private apartments, but the vast majority dine in public ; and they get but a partial view of Saratoga life, who do not scramble for a seat at the *table-d'hôte*.

In the chief hotel the dining-room is of prodigious dimensions. It is, in fact, two enormous rooms thrown into one, in the form of an L. Three rows of tables take the sweep of it from end to end. It can thus accommodate at least 600 guests. The windows of both sections of the dining-room looked into the quadrangle, and my friend and I observed that several of the loungers in the colonnade every now and then cast anxious glances within as the tables were being laid for dinner. It soon occurred to us that there might be some difficulty in getting seats, a point on which we sought to set our minds at rest, so that we might be prepared, if, necessary, for the crush. But we could effect no entrance into the dining-room to make inquiry, every approach to it being locked. At last, however, we caught in the colonnade a tall black waiter, dressed from top to toe in snow-white livery.

"Will there be any crush, when the bell rings?" I demanded of him.

"Bit of a squeeze, that's all," he replied. "But you needn't mind," he continued, regarding me, "the fat uns get the worst on't."

"Then you can't tell us where we are to sit?" said I.

"Jist where you happen to turn up, gemmen," he responded, grinning and showing his ivory.

"But surely," interposed my friend, "you can secure a couple of chairs for us?"

"It's jist within de power of possibles, gemmen," said he, grinning again, but with more significance than before. My friend slipped a quarter of a dollar into his hand. Oh! the power of money. That which was barely possible before, became not only practicable but certain in a twinkling. He immediately left us to fulfil our wishes, telling us to look in at the window and see where he secured chairs for us. The doors were still locked, but by-and-by we perceived parties of ladies and gentlemen entering the dining-room by those connecting it with the private apartments, and taking their seats at table. The *ignobile vulgus*, in the interior colonnade, were kept out until the ladies and those accompanying them were all seated. Then came the noisy jingle of the long wished-for bell. Back flew every door, and in rushed, helter-skelter, the eager crowd. We took our post at the door nearest the chairs set apart for us, on which we pounced as soon as we were pushed in, and were thus secure in the possession of places from which we could command a look of both arms of the dining-room. It was some time ere all were seated; and in the *hurry scurry* of entering it actually seemed as if some were leaping in at the windows. It was not because they were famished that they thus pressed upon each other, but because each of them wished to secure the best available seat. It was amusing to witness, as they got in, the anxious glances which they cast round the room, and then darted off in dozens for the nearest vacant chairs. At length all were seated, and the confusion subsided, but only to give rise to a new hubbub. No sooner was the signal made for a general assault upon the edibles, which were plentifully served, than such a clatter of dishes and a noise of knives and forks arose, mingled with a chorus of human voices, some commanding, others supplicating the waiters, as I had never heard before. In one room were nearly 600 people eating at once, and most of them talking at the same time. The numerous waiters were flitting to and fro like rockets, sometimes tumbling over each other, and frequently coming in very awkward collision. Every now and then a discord would be thrown into the harmony by way of a smash of crockery or crystal. The din and confusion were so terrific as utterly to indispose me to dine. I could thus devote the greater portion of my time to looking around me. The scene was truly a curious one. There were many ladies present, but the great bulk of the company consisted of the other sex. The ladies were in full dress, the *table-d'hôte* at Saratoga being on a totally different footing from that at other hotels. In about twenty minutes the hall looked somewhat like the deck of a ship after action. The survivors of the dinner still remained at table, either sipping wine or talking together, but the rest had disappeared as if they had been carried out wounded or dead. Their fate was soon revealed to us; for, on emerging shortly afterwards into the interior colonnade, we found them almost to a man seated in arm-chairs or rocking-chairs, some chewing, but the great bulk smoking. Before dinner they risked their necks to secure seats at table; after it their anxiety was to secure them on the colonnade. Hence their sudden disappearance from table.

When the day is not too hot, parties drive and walk to the springs, or to some of the most attractive points in the neighbourhood. The evenings are generally devoted to amusement, those of a public nature alternating between balls and concerts.

After spending several very pleasant days at Saratoga, I parted with my friend and proceeded by railway to Troy, a charming town with about 20,000 inhabitants, situated on the left bank of the Hudson, at the head of its navigable channel. From Troy I dropped down the river next day for seven miles by steamer to Albany. A pretty thick fog mantled the river, which the morning sun soon dissipated, and displayed to us the capital of New York, with its noble terraces and gilded domes, occupying a commanding position on the high sloping bank on our right.

So far as the trade of the West is concerned, Albany is at the head of the navigation of the Hudson. It has two highways to the sea; one the Western railway, leading due east for 200 miles to Boston; the other the Hudson, leading due south to New York. By canal, lake, and rail, it has ready access due north to Montreal; and by canal and railway the same, due west to the lakes. It is in every point of view, therefore, advantageously situated as an internal *entrepôt*, being the converging focus of four great highways, natural and artificial, from the four points of the compass. There is an upper and a lower town, the chief connexion between which is State-street, which descends the steep bank in a straight line from the capitol almost to the river. The lower town is rather crowded, chiefly owing to the narrow slip of land to which it has to accommodate itself. Albany is, on the whole, well built, and being the seat of government for the State, possesses many very showy public edifices. It is rapidly growing, its population being now about 50,000.

I was invited, on my way to New York, to spend a few days with a friend at the village of Hyde Park, about half way down the Hudson, on its left bank. To be there by an early hour in the morning, I left Albany by a steamer which left for New York late at night. It was one of the last family of boats launched upon the Hudson, and which are entirely fitted up, with the exception of the space occupied by the engines and boilers, for the accommodation of passengers. She was of prodigious length, and bore some resemblance to a great bird, with its wings expanded. Her hull was like a large board turned upon edge. As it was dark, and objects ashore scarcely visible, and as I had but a few hours to sleep, I retired to my berth soon after starting. In ascending and descending the river the boats made very brief stoppages at the intermediate towns, and to be ready to go ashore as soon as we reached Hyde Park, I but partially undressed, and threw myself on the top of the berth, with a Scotch plaid over me. Mine was the upper of two berths which occupied the state-room, or small cabin, which was one of about a hundred that led off the great saloon, the lower berth being occupied by a somewhat elderly gentleman, who had gone to bed immediately on getting aboard, and had slept soundly through the noise and hubbub of starting. I had been asleep for some time, when I was awaked by the noise of feet, rushing to and fro, directly overhead. I had no time even to conjecture the cause of it, when a tremendous crash immediately below me, accompanied by a howl and cry of terror from the old gentleman, convinced me that something dreadful had happened, or was about to do so. On looking over the edge of my berth to ascertain what had occurred, I perceived a huge rounded beam, shod with iron, and garnished with some ropes and chains, projecting for a few feet into the cabin, directly over the old gentleman's berth, to which it confined him, having, in entering, almost grated his chest. I immediately sprung into the saloon, and called for one of the stewards, by whose aid the captive was released, and just in time; for no sooner was he on his legs, ere the schooner, whose bowsprit had so inopportunely obtruded itself upon us, swung round a little, when the obtrusive bowsprit was withdrawn, tearing away many of the boards through which it had penetrated, and carrying off some of the bedclothes with it, which dropped into the water. Luckily, the old gentleman was more frightened than hurt; but so frightened was he, that, on finding himself at liberty, he bounded into the saloon in his shirt, fled as if a bulldog were pursuing him, and did not stop till he reached the other end of the huge cabin. By this time a number of ladies had popped their heads out of their stateroom doors, anxiously inquiring what was the matter, but suddenly withdrew them, in still greater alarm, on witnessing so awkward an apparition. The accident occurred near the city of Hudson, a few seconds after the steam had been shot off to enable us to halt at the town, the current drifting the boat against a schooner which was lying at anchor, and which was invisible, owing to the darkness of the night, and her neglect to carry a light. It is seldom that any steamboat accident occurs on the Hudson. Frequent as they unfortunately are in the South, particularly on the Mississippi, they are as rare on the northern waters as they are with ourselves.

In a few minutes we got disengaged from our awkward predicament, and proceeded on our way. I was so discomposed, however, by what had happened, that I thought no more of sleep; so, completing my toilette, I went upon deck, where I remained till we reached Hyde Park, at which I landed at an early hour.

I here spent three days of unmixed enjoyment with my friend and his estimable family. He was a resident of New York, where he was known and universally respected for his affability, probity, and benevolence; but he generally spent a great portion of the summer on the banks of the Hudson. He has since paid the debt which we all owe to nature. He was a native of Ireland, but had resided for about twenty years in New York. He was indefatigable in his endeavours, during their prevalence, to calm the fierce excitements engendered by the Oregon question, appealing in behalf of compromise and peace, not only to the good feeling and interests of those around him, but also to many occupying the highest stations, both in the commercial and in the political circles of this country. Men high in power

here perused these appeals, and the remonstrances
which accompanied them, nor were they without
their effect. There were few in New York held in
such universal esteem, or so favourably known to
men in high position here, in connexion with politics
and trade, as the late JACOB HARVEY.

I have known several who have sailed upon the
most vaunted of the European rivers, express an
unqualified opinion to the effect that the Hudson
was, in point of scenery, superior to them all. It is
a noble stream, both in itself and the purposes to
which it is applicable and applied. The country on
either side of it is cultivated like a garden, and from
the town of Glen's Falls to the city of New York it
is studded on either bank with a succession of cities,
towns, and villages. For some distance below Albany, its banks are comparatively tame, but by-and-
by they swell on the right, a little back from the
stream, into the majestic proportions of the Caatskill
Mountains. Hyde Park is a little below these, on
the opposite side, at a point where, in bending round,
the Hudson forms a small lake, studded with islets.
A finer view can scarcely be imagined than that obtained, looking towards the hills, from the high bank
overhanging the river at Hyde Park. I have enjoyed
it in Mr. Harvey's company, by night and by day, in
fierce sunshine, and in bright cold moonlight. The
combination of land and water, and of all that tends
to make up a magnificent landscape, is almost perfect.
The eye leaps over the intervening tree-tops, upon
the broad volume of the islet-studded Hudson, across
which it wanders to alight upon a large expanse of
undulating country, half cultivated and half wooded,
after ranging over which it reaches the foot of the

mountains, to the dreamy heights of which it then climbs. When looking at the hills, I used to amuse myself in fancying that I could pick out the spot, on their deeply wooded sides, where Rip Van Winkle slept through the Revolution.

In company with Mr. Harvey, I visited Mr. Robert Emmett, the nephew of him whom the law in Ireland, upwards of forty years ago, claimed as its victim. His home was about two miles from Hyde Park, overlooking the river at the point where, perhaps, the finest view of the prospect, just alluded to, can be obtained. We were well received by him and his lively and pretty little wife. He was both intelligent and communicative, but seemed more disposed for a quiet life than for the turmoil and strife of politics. His time was chiefly divided between his farm and his garden. His name was pretty freely mixed up with the Irish demonstrations of last year in New York; but this, I presume, was more in deference to the wishes of others, who desired to have the use of a name having no little influence with Irishmen, both at home and abroad, than from any yearning on his own part to exchange the peaceable occupations of country life for the turbulent orgies of Tammany Hall.

On leaving Hyde Park, my destination was West Point, about forty miles below; a spot possessing some interest, not only from the romantic nature of its position, and the part which it played in the revolutionary war, but also from its being now the military academy of the United States. About thirty miles below Hyde Park the stream meets with a ridge of hills known as the Highlands of the Hudson, through which it forces its way by a narrow, winding

and very romantic channel. The town of Newburg
lies upon its right bank, just above its entrance into
the gorge. About midway between Newburg and
the Tappan Zee, on the other side of the ridge, is
West Point, completely imbedded amongst the hills,
the river, at the narrowest part of its channel, sweeping round three sides of it, which gives it the command of several miles of the stream, at the most
critical point of its navigation. This point is the key
of the Hudson. It is, in fact, to the Hudson what
Rouse's Point is to Lake Champlain. It was this
important position that General Arnold was about to
deliver up to the British during the revolutionary
war, a project which was defeated by the capture of
the unfortunate André. It is, perhaps, as well for
all parties that it did not succeed, for the possession
of West Point by the imperial forces would, in all
probability, have changed the whole aspect of the war.

It is here that the future officers of the American
army are taught those branches of general and military education most befitting the career on which
they are about to enter. The establishment belongs
of course to the general government, and is under its
exclusive management and control. There is much
conflict of opinion in the United States as to the
necessity for, or the utility of, such an institution as
that at West Point. Its object is to prepare officers
cut and dry for the service; those who are in favour
of the establishment maintaining that too much attention cannot be paid to the military education of those
who may be called upon, at some future day, to lead
the armies of the United States. Others, who do not
deny the desirableness of such an education, object to
confining every post beyond the ranks in the army to

the cadets of the military academy. A private in the British army may rise to be a field officer, but not so in America. The private in the latter may be better paid than in the former, but his prospect is by no means so brilliant. There is not an office in the State, but is open to the obscurest individual, if he can beat his multitudinous competitors in the race for it. The army is not so democratically constituted. Its more desirable posts, its dignities and honours, are almost exclusively confined to a few, who have sufficient influence to get admittance to an institution, where they undergo a probationary curriculum. This is enough to discourage many a man from entering the army as a private, who might otherwise do so. If it is the policy of the American government to check the military spirit, this certainly tends to the accomplishment of its object.

Republics are accused of being prone to war. This may be partly accounted for by the citizen of a republic feeling that he participates more in the glory and honour of his country, than the subject of a monarchy, as well as feeling himself more directly involved in her quarrels. When the government is of his own creation, the position of his government in regard to a foreign power he feels to be his own. It is otherwise in a purely monarchical State, where the government is independent of, and has separate interests from those of the people. The attitude assumed by their respective courts is not necessarily that of one people towards another. The government of Russia and Austria may be at loggerheads with each other, and yet no enmity exist between the people of the two empires, except such as is created by law. But in a republic each citizen espouses the quarrel of his government as his own; and is but too ready frequently to sustain it in any project of aggression which promises to bring an accession of territory, honour, or glory to his country, and by consequence, partly to himself.

The Americans have been regarded as forming no exception, in this respect, to the general rule. But the military propensities of the American people have been very much exaggerated. They are far more ready to assume a belligerent attitude in their national, than they are to fight in their individual capacity. There is no one more ready to follow up at all hazards the fortunes of his country, or who more warmly or readily espouses his country's quarrels, than the American. He is ready to risk the chances of war, if necessary, to vindicate her honour, or to secure her a tempting prize at which she has any pretext for grasping. But all this ardour and enthusiasm resolve themselves, as a general rule, more into a willingness to submit to the national drawbacks of a state of hostility, and to give up his means and substance to maintain the war, than to subject himself personally to the privations of a campaign. How could it be otherwise in a country circumstanced as America is? Where employment is sure and wages high, men are not very willing to subject themselves to the hardships and rigid discipline of a soldier's life. The volunteers who flocked to the Mexican war were lured into the field more by the hope of realising rich prizes at the enemy's expense, than from any very great love of military adventure. At first a general enthusiasm pervaded all ranks, and it really seemed as if all were ready to buckle on their armour. But this soon subsided, and by-and-by the war grew stale.

The volunteers who did come forward, were either restless spirits from the West, to whom an adventure is a godsend, or the mere offscourings of the sea-board cities. A very large proportion of them were foreigners. Add to this that the great bulk of the American army is composed not of natives but of foreigners. The same may be said of those manning the navy. The life of an American soldier is by no means a pleasant one, considering the unhealthiness of some of their military posts, and the remoteness of many of them from the haunts of civilized man. It is not likely, therefore, that men who can easily make more than a competence at the plough or at their trades, will suffer a military propensity so far to get the better of them as to impel them to enlist.

But it may be urged that there is a great deal of sound and fury in the United States, which must surely signify something in the way of the populace being disposed to military life. It signifies very little in this way. When a dispute arises between them and another people, the Americans assume a very bellicose language, and generally, in such cases mean what they say. But this, as already intimated, does not indicate a readiness on their parts personally to take the field, draw the sword, or carry the musket. It merely testifies their readiness to run the risks of war as a people, to incur its expense and abide its issues. But, again, it may be said that the number of independent volunteer companies which are found in every part of the Union, proves that the people are, individually, prone to military life. There is a great difference, however, between " playing at soldiers " and being soldiers in earnest. To enrol themselves into a company called by some very

sounding name; to wear fine clothes, and have brilliant plumes waving over their heads; to march,
every now and then, in military array, the wonder of
a crowd of gaping boys, and the admiration of the
young ladies who present them with banners; to
undergo occasionally a review, and to engage, to
the terror of old women, in platoon firing, with
blank cartridge, in the streets,—is a pastime perhaps
harmless after all, for young men who have a
little time upon their hands, a little spare cash in
their pockets, and few other sources of amusement at
command. But all this is no proof that these valiant
men-at-arms, who generally wind up an afternoon's
marching and countermarching with a good supper
or a ball, are ready to go to the cannon's mouth, or
to abandon their peaceful pursuits for the privations
of an actual campaign. This holiday soldiering is
only, after all, but a kind of mature child's play. Let
me not be understood to mean that the Americans
are deficient in personal courage. Should their
country be invaded, none would be found more
ready to turn out and defend their altars and their
hearths. But so long as they are in comfortable circumstances at home, they will not be emulous to
take the field, unless some strong exciting cause, like
an invasion of their territory, impel them so to do.
Nor let the love of some of them for now and then
attiring themselves in military habiliments argue
anything to the contrary.

The portion of the population exhibiting to the
greatest extent the martial propensity, is that domi-
ciled in the north-west. There are many restless
spirits residing to the north and north-west of the
Ohio, so fond of adventure, that they will, in most

cases, undergo any personal risks in pursuit of it. When, in addition to this innate love of adventure for its own sake, a great prize is presented to them, the securing of which will enure to their individual advantage and to the glorification of the Union, they are ready to leave home and friends to grasp at it. But by-and-by, when this part of the country is more advanced, and property in it becomes more valuable, rendering a permanent settlement in it, a thing once obtained not to be lightly thrown away, this restlessness will greatly disappear, and the people sober down to the tone of the great bulk of their countrymen. Besides, there is this also to account for the West being more reckless of war than the other sections of the country, that unless the people there chose to subject themselves to them, they would be the last to feel its privations. The Union is vulnerable on three sides, but the valley of the Mississippi would be secure from the horrors of war, should it arise.

Until within a few years back the United States army did not exceed 8,000 men. It was found, however, that as the Republic extended its boundaries and multiplied its military posts in the remote wildernesses which circumscribe it on the west and north-west, this number did not suffice to garrison and keep in repair the more important military stations, scattered at long intervals along its extensive frontier. The standing army was therefore increased about seven years ago to 12,000 men, at which point it remained till the breaking out of the Mexican war. It was then necessarily increased, but for the year 1848, which witnessed the successful close of the war, it did not exceed 25,000 men. Of the

American forces which took part in the Mexican campaigns, the volunteers formed a large and important ingredient.

The American navy in 1848 was on an equally limited scale, although the war lasted till about the middle of that year. The total number of vessels of all kinds connected with it in November last amounted to eighty-seven; of which eleven were ships of the line, fourteen were frigates, twenty-two were sloops of war, ten were schooners, and fourteen were steamers. The war did not occasion a similar increase in the navy to that called for in the army, inasmuch as the Mexicans had no navy to cope with; at the same time that, to their honour, they refrained from issuing letters of marque. This naval force suffices for the protection of American commerce, which, if not as yet absolutely as large as our own, spreads, in its multiform operations, over an equally extensive surface.

There are, undoubtedly, interests at work in the United States which would benefit, as in this country, by the indefinite extension of the military establishments. But mighty armaments, particularly in the form of land forces, would be incompatible with the objects and inimical to the very genius of the American constitution. The government was conceived in the spirit of peace, and framed more with a view to aid and encourage the development of the peaceful arts, than to promote a martial spirit in the people, or to throw the destinies of the country into a military channel. Not only do the views, sentiments, and occupations of the American people indispose them to any great permanent increase of the military establishments, but there are, as I found, conflicting elements at work in different sections of the Confederacy, which would of themselves suffice to confine them to moderate limits. Whilst it is the object of the sea-board States, in which the chief commercial and shipping interests of the Union are centred, to increase the navy, the West is averse to its extension, having no direct interests to subserve by its increase. The West, on the other hand, would have no great objection to some increase in the army, but the sea-board States, having little or nothing to gain from such a step, are averse to its being taken. Thus, between their conflicting views and wishes, the establishments, except in cases of extraordinary emergency, are not likely to receive any very great accession of strength. This at all events may be said, that no accession will be made to either of them until a clear case of necessity for it is made out. The average annual cost of the United States navy for the last ten years has not exceeded 1,295,000*l*. The average annual cost of the army for the same period has been about 2,500,000*l*., but this includes not only the extraordinary military expenditure occasioned for some years by the Seminole war, but also a portion of that called for during the first year of the Mexican war. Making due allowance for this extraordinary expenditure, the average cost per annum of the army will not exceed 1,500,000*l*. Taking the two services together, their average cost per year is thus shown to be about 2,795,000*l*.; about one-sixth of the sum which we are now called upon to pay for our armaments.

It may be urged that the great reason why the American establishments are kept at so low a point is, that the military exigencies of the country are not so great as they formerly were. It is quite true that, as the Republic has extended itself, its military boundary, in the strict sense of the term, instead of increasing, has diminished. At the peace of 1783 it was enclosed on three sides by the dependencies of foreign powers. The British provinces overhung it on the north, the vast French possession of Louisiana spread along its entire western boundary, and the colonies of Spain underlay it, as it were, on the south. Since that time it has acquired Louisiana from France and the Floridas from Spain, and has recently pushed its boundaries westward to the Pacific. Its land boundaries are now confined to the line separating it from the British possessions on the north, and that which divides it from what now remains of Mexico, to the south. But with the diminution of its land boundary its sea coast has greatly increased. At the peace its only sea-board was on the Atlantic, stretching from the Bay of Fundy to the mouth of the St. Mary's, which separated it from Florida. It afterwards crept round the immense peninsula of that name, and along the northern shore of the Gulf of Mexico, beyond the mouths of the Appalachicola, the Alabama, and the Mississippi, to that of the Sabine. Thence it proceeded westward to the Nueces, and lastly to the Rio Grande itself, the left bank of which now forms its south-western boundary. It thus gradually possessed itself of the Atlantic and Gulf shores of Florida, the whole north coast, and the north-western angle of the Gulf. In addition to this, it has lately acquired an immense stretch of sea-board, from the Straits of Fuca, the northern point of American Oregon, to the southern limit of Upper California. But with this immense accession of sea coast the American navy has shown no greater tendency to extension than has the army. Indeed, the chief extension which has taken place has been in connexion with the latter, for although the land line has diminished as that of the sea-board has increased, civilisation in its rapid spread westward has required, for its protection against the Indians, that a more efficient military cordon should be kept in advance of it, than was necessary when it was confined to the eastward of the Mississippi.

It is not easy to estimate the effect which these enormous accessions of sea-board are calculated to have upon the naval resources of America. This much, however, may be safely taken for granted, that the increase with which they have been attended, to the naval strength of the country, has not been commensurate with their own extent. As compared with the American line of coast, the British American available sea-board is small. But its importance in a naval point of view is as it were in the inverse ratio of its extent. Both on the Atlantic and the Pacific the British flag yet waves over the most important harbours in a military point of view, and over the most commanding line of sea-board. The British available sea-board on both sides of the continent is not, in extent, more than one-fourth the whole sea-coast of the Union, yet the possession of it would at once treble the naval strength of the Union. Not only are Bermuda, Halifax, and the mouth of the St. Lawrence military stations of the highest importance, but the possession of our North American provinces would put the finest fisheries in the world into the hands of our rivals.

In perusing these paragraphs the reader cannot fail

to be struck with the contrast which they present between our own military establishments and those of
the United States.   It may very truly be urged, that
the military exigencies of Europe are not to be measured by those of America.   But although there is,
in this respect, a great difference between America
and a continental State, the difference is not so great
between the United States and Great Britain. Thanks
to their isolation from Europe, the Americans are
under no necessity to keep large and expensive military establishments on foot.   But are not we also
isolated from Europe?   We are nearer it, it is true,
but our isolation from it is as complete as is that of
the United States.   The immense advantage which
this gave us, we have not only trifled with, but
thrown away.   Since the " balance of power " came
to be a leading and favourite notion with European diplomatists, we have needlessly mixed ourselves up with every great and every petty squabble
that has happened on the continent.   The result has
been as unfavourable to us as if the channel had
been dried up, and we had been long ago geogra-
phically annexed to the continent.   We have unnecessarily worked ourselves into a position which
we might easily have avoided, and from which, it
must be confessed, it is not now easy to recede, even
were we unanimous as to the propriety of so doing.
But instead of lessening our difficulties in this respect, and taking all the economic advantages of our
position which it is calculated to confer upon us, we
are involving ourselves every year more and more
in the vortex of continental politics, and are consequently called upon to increase rather than to
diminish our armaments.   With Sicily and Lom-

bardy, Rome and the "two duchies" to take care of, to say nothing of our quarrel with Spain, and our recent interference with the Portuguese, the prospect before us is not very encouraging. We have two courses to pursue, either to go on systematically intermeddling with affairs in which we are not necessarily concerned, until we concern ourselves with them, thus subjecting ourselves to the military necessities of a continental State; or to relapse as it were into ourselves, devote our attention as exclusively as possible to our home and colonial affairs, take advantage of our defensive position, conform our military establishments to the measure of our strict wants, and curtail our extravagance. It is not necessarily England's mission to undertake the quixotic enterprise of keeping the world right. In attempting it heretofore, if she has not received many cuffs and bruises, she has at least had to submit to enormous abstractions of her treasure. Let her keep herself right, and in the industrious, peaceable attitude which she will then assume, she will do far more towards tranquillising the continent, than by vexatiously interfering in every political movement that occurs.

But the most plausible excuse offered for the greatness of our military establishments is the vast extent of our colonial dominions. As to how far every one of the forty colonies or so which we possess is of use to us, is a question into which I have here no intention of entering. It may be said, however, in passing, that some of the finest of them are comparatively useless to us, simply because the colonial department either cannot or will not turn them to profitable account. The only point with which we have here to deal is, whether the excuse alluded to is a valid one or not,

If we are to have colonies, nobody can reasonably grudge whatever is necessary for their protection. But the question is, what is necessary for this purpose? It would seem that, in the estimation of one class at least in this country, a department cannot be efficient unless it is extravagant, although daily experience teaches the contrary, some of our most extravagant being amongst our least efficient departments. The colonial department has, within the last seventy years, undergone in this respect a modification for the worse. Previously to the American war, without leaving the colonies unnecessarily exposed, we taught them the useful lesson of self-reliance. The consequence was that, until we attempted to avert, in 1776, an irresistible event, some of the noblest colonies that we ever possessed cost us but little either to govern or to defend them. Nor were these colonies wanting in formidable enemies, against whom they had to be on their guard. They had at first the fierce and cunning Indian to cope with, and were afterwards hemmed in on three sides by France and Spain, who had the Indians frequently in league with them. Against all these they, in the main, defended themselves, sometimes coping single-handed with their enemies, and at others forming leagues, the germs of the future Union, for their common defence. Having thus to bear the brunt of the fight, and the chief expenses of the war when it arose, they were chary of getting into quarrels with their neighbours, their interests being identified with peace. But this policy, at once so useful to the colonies and convenient to the mother country, was afterwards abandoned, and another inaugurated in its stead, the practical operation of which is to keep the colonies as much and as long as possible in leading-strings, and the tendency if not the object of which is to destroy in them every principle of self-reliance. We teach them that almost everything will be done for them by us, and at our expense. We will* govern them at our expense, and if they get into quarrels we will work them out of them at our expense. The consequence is, that governing them at our expense gives us a pretext for vexatiously interfering in the conduct of their local government; whilst, by protecting them at our expense, we make it their interest, in many cases, to get into quarrels with their neighbours instead of remaining at peace with them. One can understand how it would subserve the interests of Cape Town that the colony of the Cape should be at war with the Kaffirs for the next half century, so long as British regiments were sent there to spend British money in the colony, and the commissariat was supplied at the expense of the mother country. If we want to hear little of Kaffir wars, let us put the Cape colony on the footing that was formerly occupied by our dependencies in North America.

Besides, if it is simply for their protection that we keep such large armaments in and about our colonies, how comes it that the more populous they are, the stronger they become, and consequently the more competent to protect themselves, the more troops do we pour into them? Is not this of itself the most damning commentary that can be offered on the spirit in which our whole colonial system is conceived? The truth is, that we send additional troops to them, in order to enable us, as they wax stronger, to continue the vexatious interference in their local affairs, in which we so unwisely persevere.

Our peace establishment in Canada amounts to about 6,000 men. We have, in addition to this, a large naval force on the lakes, and of course an expensive commissariat for the supply of both services. Wherefore, at present, all this display in Canada? By what foe is it menaced? It has no Indian enemy against which now to protect itself. Do we apprehend an attack upon it from the side of the United States? Such cannot be effected in a night, and wars are not now declared in a day. If the Americans meditated an attack, they would have to arm for the purpose, for there is but a small portion of their regular army on the Canadian frontier. Their militia system is universal, but it is confessedly inefficient. Whilst they were arming, what would prevent Canada from arming likewise? The Canadians are more of a military people than the Americans, and in Upper Canada particularly there are elements out of which a strong military force could be more speedily evoked than out of those existing on the American side of the line. Besides, when the Americans were arming, what would prevent us from sending troops to the scene of danger? They would get there quite as soon as a force could be raised in New York. If we have 6,000 men there now for the defence of Canada, we have more than we require. If they are there to keep the Canadians down, we have less than we require; for such are the means of passive resistance at their disposal, that, in case of a general insurrection, 60,000 would not suffice to suppress it. For which purpose are they there? If for the one, the means are inadequate to the end; if for the other, the end is as questionable as the means are insufficient.

CHAPTER VIII.

EDUCATION AND LITERATURE IN THE UNITED STATES.

THE tourist may spend a very pleasant day or two, rambling over West Point and its neighbourhood. I left on the morning after my arrival at it, and in half an hour after quitting the wharf, having emerged from the highlands, found myself on the noble estuary of the Hudson, already alluded to as the Tappan Zee. New York was still forty miles distant; but from the lofty paddle-box I could discern the smoke of the city sullying the horizon to the south. The day was bright and clear; every object on either shore, notwithstanding the great width of the river at this point, being visible to us. On our left we passed Sing Sing, the other State Prison, or Penitentiary, of New York, and the mouth of the Croton, a portion of whose limpid waters, as has been shown, are diverted for the supply of the city. We soon had the " Palisades " on our right, the New Jersey coast of the river being here lofty, bold, and precipitous; masses of rock, apparently of basaltic formation, overhanging the water in columnar grandeur. The New Jersey coast on the west continues nearly up to the highlands, whence, upwards, the river is exclusively within the limits of New York. The portion of that State which continues along its eastern side down to the city, is a perfect contrast to the bold, rocky bank opposite. The New York bank is lofty, but it rises gently, with undulations, from the margin of the stream. The great extent of surface which it thus exposes is beautifully cultivated, and dotted with mansions and farmhouses.

Once more in New York, which presents the same busy and stirring picture of impetuous life as before. Having already, however, sufficiently described the city, I shall not delay the reader with any notice of my second visit. I prefer, and so, no doubt, will he, that we should sail together up the Sound to New Haven, in the State of Connecticut.

The site of this town is very picturesque. Although not very populous, it presents, from the water, the appearance of a large city, from the great length to which it extends along the shore of the open bay, entering from the Sound, on which it is situated. As a place of residence there are few spots more inviting than New Haven. It looked to me like a town spending the summer months in the country. It is scattered over a great surface, the streets being broad and spacious, and deeply shaded by rows of the most stately elms. But that which gives to New Haven its chief interest is its being the seat of the principal University in the United States. Yale College was founded at the commencement of the eighteenth century, but was not removed to New Haven until seventeen years after its foundation, where it has since been permanently retained.

There is much in the general polity of America to strike the stranger with surprise, but nothing more calculated to excite his admiration, than the earnestness with which education is there universally promoted by the State, as a matter in which the State has the most deep and lasting interest. The American government is one which shrinks not from investigation, but covets the intelligent scrutiny of all who are subjected to it. It is founded neither on force nor fraud, and seeks not, therefore, to ally itself with ignorance. Based upon the principle of right and justice, it seeks to league itself with intelligence and virtue. Its roots lie deep in the popular will, and in the popular sympathies is the chief source of its strength. It is its great object, therefore, to have that will controlled and those sympathies regulated by an enlightened judgment. It thus calls education to its aid, instead of treating it as its foe.

Let those, who will, deny that the tendencies of human nature are to good, this is the broad principle upon which the American system of government rests. There is a great difference between believing in the better impulses of our common nature, and cherishing an " idolatrous enthusiasm" for humanity. The founders of the American system kept the brighter side of human nature in view when they organized their polity, instead of acting chiefly with a view to its darker traits. They did not lose sight of the propensity to evil, which so universally finds a place in the divided heart of man, but they framed their system more with a view to the encouragement of virtue than the repression of vice. They had no blind faith in the supremacy of good over evil in the moral nature of man, but they acted throughout upon the conviction that man's social and political condition had much to do, although not every thing, with the development of his moral character. The tendency to good may be cherished, the propensity to evil checked, by the position which a man is made to occupy with regard to his fellows. A man's moral nature is not only evidenced, but also greatly influenced by his acts. Place him in a position in which the temptations to evil are more potent than the stimulants to good, and if he give way, his consequent familiarity with evil acts increases the propensity to them. But surround him with better influences, and every time he yields to them he strengthens the higher impulses of his nature. A man's conduct is thus not only the result of his moral character, but it also, to some extent, influences it. And what chiefly influences his conduct? The circumstances in which he is placed. The great object of philanthropy and of sound policy in the government of mankind should therefore be to mould these circumstances so as to stimulate to good, instead of being provocative to evil. This was the great object after which the noble race of men, who framed the American Constitution, honestly and earnestly strained. They repudiated a system founded upon the principles of suspicion and resistance, and adopted one based upon those of confidence and encouragement. Faith in, not idolatry of, human nature was thus at the very foundation of the edifice which they reared; and they took care, in arranging the superstructure, that that in which they trusted, the tendency to good—which, however it may

be sometimes smothered in the individual, can never
be obliterated from the heart of man—should have
every opportunity given it of justifying their confidence. The sympathies of ignorance are more with
the evil than with the better principle of our com-
posite natures; and they made it a primary object
of their policy to assail ignorance, in every form
in which it presented itself. The sympathies of
intelligence, on the other hand, are more with
virtue than vice; and the universal promotion of
education was made one of the main features of their
governmental system. They thus regarded education
in its true light, not merely as something which
should not be neglected, but as an indispensable coadjutor in the work of consolidating and promoting
their scheme. They had not only cause to further
education, but they had every reason to dread ignorance. They have so still, and the institutions of
America will only be permanently consolidated, when
intelligence, in a high stage of development, is homo-
geneous to the Union. The American government,
founded upon the principle of mutual confidence,
thus wisely takes care that education shall be promoted, as one of the essential conditions to the realisa-
tion of its hopes. Its success is thus identified with
human elevation—it can only be defeated by the
degradation of humanity.

How different is a system thus conceived from that
propounded by statesmen, who preach, as the fundamental element of good government, a distrust of the
moral attributes of man! They admit that he has
some good in him, but insist that he should be
treated, both socially and politically, on the supposition that the propensity to evil was the only charac-

teristic of his nature. Whether it be originally his
chief characteristic or not, there is no doubt but that
it may be artificially made so, and systems of government founded on deception, hypocrisy, and selfishness,
can never be made the means of purifying the heart,
elevating the sentiments, or exalting the intellect of
mankind. Thoroughly to improve a people, you
must, as in the case of an individual, appeal to their
generous sentiments. But a government turns its
back upon these, which shows, in the very principle
of its being, and in its every act, that it deals with the
people on the footing of distrust. It is not by the
repressive system that vice can be most effectually
eradicated. It is by promoting the antagonist principle of virtue that the greatest victory is to be achieved
over it. Systems chiefly, if not exclusively, framed
for the suppression of vice, are not the best calculated
for the promotion of virtue.

Again, systems prominently embodying the principle
of resistance, provoke resistance. The result is a
chronic antagonism between the government and the
governed, whereas harmony between the two is at once
the essence and the symbol of good government. The
principle of resistance has been tried and found wanting. Men cannot be permanently governed through
force and fear. They may be so through the affections,
and this without idolizing humanity. Force and fear
have failed; and those who relied upon them blame
humanity for their failure. May it not be that
it is a very hopeful feature in humanity that they
have not succeeded? Resistance is still preached
as the fundamental element of good government, by
one who affords in his own person the most memorable
modern example of the utter fallacy of such a prin-

ciple. It was only in 1830, whilst a spectator of the revolution of that year, that M. Guizot really learnt what were the essential elements of human society, and the indispensable prerequisites to safe and efficient government. After having imbibed this great lesson, he was for eleven years a minister. How much he profited by it, the events of February can attest. These events are the best illustration, both of the soundness of his judgment and the correctness of his system. Either he read the human heart aright in 1830, and afterwards governed his fellow-subjects on wrong principles, or he was egregiously at fault both in reading and governing them. But it was not king Louis Philippe's system, which received its chief manifestation during the seven years for which M. Guizot was virtually the head of the Cabinet, that was faulty, it was the vile human heart. There was nothing incompatible with the dignity and stability of a government in the broken faith pledged at Eu; in the despicable intrigue of the Spanish marriages; in the double dealing with the Sonderbund; in the coquetting with Colletti; in the evident leaning to the principle of despotism, typified by the rupture with England, and the growing alliance with the absolutist powers; or in the unequivocal determination to check the progress of rational liberty in France, and to suppress every noble aspiration in which she indulged. These are the leading features of the Guizot administration. Were they such as to recommend it to an ardent people, who worshipped at least the semblance of freedom, if they did not rightly appreciate its meaning? Let this be its commentary. On the 22d of February, the minister was in the plenitude of his power, the dynasty in possession of

France. A few days afterwards, and Lamartine was in the Hôtel de Ville, Louis Philippe at Claremont, and M. Guizot once more at Ghent. But it was the vile human heart that did it all. France was both insensible and ungrateful. So insinuates the fallen minister now.

The reader will pardon this digression. But, in considering the principles which lie at the foundation of the American system, I could not avoid contrasting them with certain views as to the proper elements of good government, which have recently emanated from a distinguished source.

I have already intimated that the American government, instead of seeking to fortify itself in popular ignorance, and to make society virtuous by simply resisting the propensity to evil, is framed with a view to strengthen and encourage the tendencies to good —the possession of which, to some extent, even his greatest detractors cannot deny to man—and allies itself with education as its most potent coadjutor in the work. It has already been seen that the general government is but a part of what is understood by the political system of America; and that the State governments form its main, if not its most interesting feature. In speaking of the close alliance formed between the American system and general education, let me be understood to refer to the system in its local, not its federal manifestation. The education of the people is not one of the subjects, the control over which has been conceded to the general government. There were two reasons why the different States reserved its management to themselves. The first was the difficulty of procuring a general fund for its support, without investing the general government with some power of local taxation, a course which would have been at war with some of the fundamental axioms of the whole system. The other was the impossibility of devising a general plan of education for a people, whose political being was characterised by so many diversities of circumstances, and who differed so essentially from each other in some of their institutions. The States, therefore, prudently reserved the management of the whole subject to themselves. The cause of education has not lost by this; the States, particularly those in the north, running with each other a race of generous emulation in their separate efforts to promote it.

In a country in which the Church has been wholly divorced from the State, it was to be expected that education would be divested of the pernicious trammels of sectarian influence. The Americans have drawn a proper distinction between secular and religious instruction, confining the Church to its own duties, and leaving the schools free in the execution of theirs. They have not fallen into the ridiculous error of supposing that education is " Godless," when it does not embrace theology. Education has both its secular and its religious elements. As men cannot agree as to the latter, let not the former, on which they are agreed, be prevented from expanding by unnecessarily combining them. Cannot a mathematical axiom be taught, without incorporating with it a theological dogma? Is it necessary, in order to rescue this branch of education from the charge of godlessness, that a child should be taught that it is with God's blessing that the three angles of a triangle are together equal to two right angles; or that two and two, *Deo volente*, make four, otherwise they might have made five ? Suppose, then, that we had schools for teaching arithmetic and mathematics alone, would any sane man charge them with being godless because they confined themselves to the teaching of such simple truths as that two and two make four, and that the three angles of a triangle are together equal to two right angles ? And what holds good of a branch of secular education, holds good of it in its entirety. If mathematics can be taught without theology, so can reading and writing, grammar and geography ; in short, every department of secular learning. This is the view which the Americans have generally taken of the subject, and they have shaped their course accordingly. They have left religion to fortify itself exclusively in the heart of man, whilst they have treated secular education as a matter which essentially concerned the State.   Either the Church is fit for the performance of its own duties, or it is not.   If it is not, it is high time that it were remodelled ; if it is, there is no reason why it should call upon the school to undertake a part of its work.   The school might, with the same propriety, call upon the Church to aid it in the work of secular instruction.   They will both best acquit themselves of their responsibilities, when they are confined exclusively to their own spheres.   In America they are so, and with the happiest results. The children of all denominations meet peaceably together, to learn the elements of a good ordinary education.   Nobody dreams of their being rendered godless by the process.   Their parents feel assured that, for their religious education, they can entrust them to the Church and the Sunday-school.

The importance which the American people attach

to the subject of general education, is indicated by the prominent position which they assign it, amongst those matters which peculiarly claim the attention and supervision of the State. As is the case in some of the States of the continent, in most, if not in all, of the States of the American Union, the superintendence of education is made a separate and distinct department of State. He who presides over this department may not be permitted to appropriate to himself so high-sounding a title as Minister of Public Instruction; but nevertheless, within his own State, he is such minister. We manage things differently. We have no separate department for the supervision of this all-important subject. We have the Home department, whose chief business it is to war with vice, and to preserve the public peace against those who would be disposed to break it. This is very necessary to the existence of society. But there is no department to carry on the war with ignorance, and to dispose to virtue by enlightening the mind. This noble object is almost exclusively entrusted to a Committee of Privy Council, who delegate their duties to a single individual, who, however responsible he may be to his employers, is not directly responsible to parliament. It is quite possible that this committee may answer all the purposes of a separate department, but it is not probable that it does so. The spirit in which all our national schemes for the education of the people are conceived, is evident from the very nature of the superintendence to which they are subjected. Education is regarded by our rulers as a subsidiary matter, or its charge would not be committed to a species of irresponsible committee. This neglect of or apparent contempt for, education, on the part of the government, has a pernicious effect upon multitudes in the country, who only permit those things to rank high in their estimation which are treated by government with dignity and respect. Let the government once elevate education to its right position, as one of the primary objects of State solicitude, and let its supervision be entrusted to parties directly responsible to parliament, and numbers without, who are now indifferent to the subject, would zealously co-operate in its promotion. Let this be done—let the superintendence of education be organized into a distinct department of the government, and we should not much longer have to blush at the scandal of the yearly expense of education in this great country turning up as a paltry item in the miscellaneous estimates.

Nothing can better serve to illustrate the difference of spirit, in which our educational system and that of America are conceived than the yearly outlay by the State in both cases, in the way of its promotion, as compared with other items of national expenditure. We pay nearly nine millions a-year for the support of one only of our military establishments, and about 130,000*l.* for popular education ; whereas, the largest item in the annual expenditure of several of the States of the Union, such as Connecticut and Rhode Island, is for the promotion of the education of the people.

The States of the Union differ not only in the form of their educational schemes, but likewise in the extent to which they have pushed them. It is in the northern States that the noblest efforts have been made for the spread of popular instruction. In the slave-holding States such schemes as have been adopted, have been rendered applicable only to the white population. But with this solitary exception there is not a State in the Union that has not done something, and most of them a very great deal, for the promotion of popular education. It would not be advisable here to enter into the details of their different schemes, but those of one or two States may be briefly glanced at by way of illustration.

All the New England States have done much in this behalf. That which has been effected by Connecticut, will show the spirit in which the great work has been taken up by the Americans in their political capacity.

The population of this State does not exceed that of the city of Glasgow. It has a permanent school-fund, amounting to about two millions of dollars, or 416,666*l.* sterling. This yields an annual revenue of about 120,000 dollars, or about 25,000*l.* sterling. The fund, I understand, has lately increased, the revenue which it yields being now about 26,000*l.* The State is divided into upwards of 1,660 school districts, in all of which schools are in operation. In 1847, upwards of 80,000 children were instructed in all the elements of a good ordinary education at these schools; the rate per child, at which they were taught for a year, being 1 dollar and 45 cents, or about 6*s.* sterling. In addition to this, there are in the State several colleges, and upwards of 130 academies and grammar-schools, the State confining its operations to the bringing home to every citizen a good elementary education. And it is only when the State as a State undertakes the work, that it can be done in the effectual manner in which it has been achieved in Connecticut. Our annual State expenditure on education is a little over 100,000*l.* Were our expenditure in this respect on the same scale, in proportion to our population, as that of Connecticut, instead of 100,000*l.* it would be

2,288,000*l*., or nearly twenty times as great as it is. But, as regards the provision which she has thus made for education, Connecticut stands preeminent even in America.

The State of New York has also set a noble example, in this respect, to the other communities of the world. The population of this State is under three millions. It is divided for the purposes of education into school districts, which constitute the lowest municipal subdivisions of the State. The number of these districts is 10,893! In 1843 schools were open in no less than 10,645 of these. The number of children from five to sixteen years old in these districts was 601,766. Of these no less than 571,130 were attending school. Upwards of half a million of dollars was, that year, paid to teachers by the State. The whole amount paid by the State for education in 1846 was 456,970 dollars, or 95,202*l*. sterling ; and this for the education of between two millions and a half and three millions of people. If we spent at the same rate for the same purpose, our yearly expenditure for education would be 1,142,424*l*., or very nearly ten times as great as it is. It is quite true that enormous sums are voluntarily appropriated in this country to the purposes of education. But it would be erroneous to suppose that this is not also the case in America, where such large sums are annually expended upon education by the State. In addition to the common schools, of which all who choose may avail themselves, and in which a sound elementary education alone is taught, there are in New York nearly 600 academies and grammar-schools, which do not enter into the State system at all, and at which the higher branches of education are taught. New York also abounds in seminaries of the highest grade, chief amongst which are Columbia College and New York University, both in the city of New York, and Union College in the city of Schenectady.

Let it not be supposed that because the common and primary schools have been rescued from sectarian influence, the different sects in the country have no educational institutions of their own. They have none designed to supersede the primary schools, such as they possess being institutions to which youth resort only when they leave these schools. Although not all, most of the colleges in America are of this description. Of 109 colleges in the United States, 10 are institutions belonging to the Baptists, 7 are Episcopalian, 13 are Methodist, and several Catholic. The great bulk of them seem to be divided between the Congregationalists and Presbyterians, the former possessing most of those which are in New England, and the latter the majority of such as are scattered throughout the rest of the country. There are also 35 theological schools in the country, of which 6 are Congregationalist, 11 Presbyterian, 3 Episcopal, and 5 Baptist. Law and medical schools are likewise numerous throughout the Union.

The number and magnitude of the seminaries existing in the State of New York for the education of young ladies form a striking feature in the educational system of that State. Most of the pupils at these establishments are boarders, and their education generally takes a much wider scope than does that of young ladies in this country. Their scientific acquirements are, however, attained at the expense of their accomplishments.

The results of the general attention to popular education characteristic of American polity, are as cheering as they are obvious. It divorces man from the dominion of his mere instincts, in a country, the institutions of which rely for their maintenance upon the enlightened judgments of the public. Events may occur which may catch the multitude in an unthinking humour, and carry it away with them, or which may blind the judgment by flattering appeals to the passions of the populace; but on the great majority of questions of a social and political import which arise, every citizen is found to entertain an intelligent opinion. He may be wrong in his views, but he can always offer you reasons for them. In this, how favourably does he contrast with the unreasoning and ignorant multitudes in other lands! All Americans read and write. Such children and adults as are found incapable of doing either, are emigrants from some of the less favoured regions of the older hemisphere, where popular ignorance is but too frequently regarded as the best guarantee for the stability of political systems.

In a country of whose people it may be said that they all read, it is but natural that we should look for a national literature. For this we do not look in vain to America. Like its commerce, its literature is as yet comparatively young, but like it in its development it has been rapid and progressive. There is scarcely a department of literature in which the Americans do not now occupy a respectable and prominent position. The branch in which they have least excelled, perhaps, is the drama. In poetry they have been prolific, notwithstanding the practical nature of their pursuits as a people. A great deal of what they have produced in this form is valueless, to say nothing else; but some of their poets have deservedly a reputation extending far beyond their country's bounds. Of the novels of

Cooper it is not necessary here to speak. There is an originality in the production of Pierpoint, and a vigour in those of Halleck, a truthfulness as well as force in the verses of Duna, and a soothing influence in the sweet strains of Bryant, which recommend them to all speaking or reading the glorious language in which they are written. In the bright galaxy of historical authors, no names stand higher than do those of some of the American historians. The fame of Prescott has already spread, even beyond the wide limits of Anglo-Saxon-dom. The name of Bancroft is as widely and as favourably known; his history of the United States, of which only a portion has as yet appeared, combining the interest of a romance with fidelity to sober realities. In biographical literature, and in essays of a sketchy character, none can excel Washington Irving; whilst in descriptive writing, and in detailing " incidents of travel," Stevens has certainly no superior. Many medical works of great eminence are from American pens; and there is not a good law library in this country but is indebted for some of its most valuable treasures to the jurisprudential literature of America. Prominent amongst the names which English as well as American lawyers revere, is that of Mr. Justice Story. Nor have American theologians been idle, whilst jurists and physicians have been busy with their pens. Dwight, Edwards, and Barnes, are known elsewhere as well as in America as eminent controversialists. Nor is the country behind in regard to science, for not only have many valuable scientific discoveries been made and problems solved in it, but many useful works of a scientific character have appeared, to say nothing of the periodicals which are conducted in the interest of science. The important science of Economy has also been illustrated and promoted by the works of American economists, whilst Americans have likewise contributed their share to the political and philological literature of the world. The American brain is as active as American hands are busy. It has already produced a literature far above mediocrity, a literature which will be greatly extended, diversified, and enriched, as by the greater spread of wealth the classes who can most conveniently devote themselves to its pursuit increase.

It is but natural that a government which does so much for the promotion of education should seek to make an ally of literature. Literary men in America, like literary men in France, have the avenue of political preferment much more accessible to them than literary men in England. There is in this respect, however, this difference between France and America, that whilst in the former the literary man is simply left to push his way to place; in the latter, he is very often sought for and dragged into it. In France he must combine the violent partisan with the literateur ere he realises a position in connexion with his government. In America the literateur is frequently converted into the politician, without ever having been the mere partisan. It was thus that Paulding was placed by President Van Buren at the head of the navy department, that Washington Irving was sent as minister to Spain, and Stevens despatched on a political mission to Central America. It was chiefly on account of his literary qualities that Mr. Everett was sent as minister to London, and that Mr. Bancroft was also sent thither by the cabinet of Mr. Polk. Like Paulding, this last-mentioned gentleman was for some time at the head of a department in Washington, previously to his undertaking the embassy to London. The historian exhibited administrative capacity, as soon as he was called upon to exercise it; whilst in this country he has earned for himself the character of an accomplished diplomatist, a finished scholar, and a perfect gentleman. But Mr. Bancroft's future fame will not depend upon his proved aptitude for administration or diplomacy. As in Mr. Macaulay's case, so with him, the historian will eclipse the politician.

As is the case in this country, the periodical and newspaper press occupies a very prominent position in the literature of America. Periodicals, that is to say, quarterlies, monthlies, and serials of all kinds, issue from it in abundance; the reviews and magazines being chiefly confined to Boston, Philadelphia, and New York.

In connexion with American newspapers, the first thing that strikes the stranger is their extraordinary number. They meet him at every turn, of all sizes, shapes, characters, prices, and appellations. On board the steamer and on the rail, in the counting-house and the hotel, in the street and in the private dwelling, in the crowded thoroughfare and in the remotest rural district, he is ever sure of finding the newspaper. There are daily, tri-weekly, bi-weekly, and weekly papers, as with us; papers purely political, others of a literary cast, and others again simply professional; whilst there are many of no particular character, combining every thing in their columns. The proportion of daily papers is enormous. Almost every town, down to communities of two thousand in number, has not only one but several daily papers. The city of Rochester, for instance, with a population a little exceeding 30,000, has five;

VOL. III.   M

to say nothing of the bi-weekly and weekly papers which are issued in it. I was at first, with nothing but my European experience to guide me, at a loss to understand how they were all supported. But I found that, in addition to the extent of their advertising patronage, which is very great, advertisements being free of duty in America, the number of their readers is almost co-extensive with that of the population. There are few in America who do not both take in and read their newspapers. English newspapers are, in the first place, read but by a few; and in the next, the number of papers read is small in comparison with the number that read them. The chief circulation of English papers is in exchanges, news-rooms, reading-rooms, hotels, taverns, coffeehouses, and pot-houses, but a fraction of those who read them taking them in for themselves. Their high price may have much to do with this. In America the case is totally different. Not only are places of public resort well supplied with the journals of the day, but most families take in their paper, or papers. With us it is chiefly the inhabitants of towns that read the journals; in America the vast body of the rural population peruse them with the same avidity and universality as do their brethren in the towns. Were it otherwise it would be impossible for the number, which now appear, to exist. But as newspapers are multiplied, so are readers, every one reading and most subscribing to a newspaper. Many families, even in the rural districts, are not contented with one, but must have two or more, adding some metropolitan paper to the one or two local papers to which they subscribe.

The character of the American press is, in many

points of view, not as elevated as it might be. But in this respect it is rapidly improving, and, as compared to what it was some years ago, there is now a marked change in it for the better. There may be as much violence, but there is less scurrility than heretofore in its columns; it is also rapidly improving in a literary point of view. There are several journals in some of the great metropolitan cities, which, whether we take into account the ability with which they are conducted, or the dignity of attitude which they assume, as favourably contrast with the great bulk of the American press, as do the best conducted journals of this country.

The American papers, particularly in the larger commercial towns, are conducted with great spirit; but they spend far more money in the pursuit of news than they do in the employment of talent. Their great object is to anticipate each other in the publication of news. For this purpose they will either individually, or sometimes in combination, go to great trouble and expense. During the progress of the Oregon controversy a few of the papers in New York and Philadelphia clubbed together to express the European news from Halifax to New York, by horse-express and steamer, a distance of 700 miles, and this too in winter. The most striking instance of competition between them that ever came under my observation was the following. For some time after the breaking out of the Mexican war, the anxiety to obtain news from the South was intense. There was then no electric telegraph south of Washington, the news had therefore to come to that city from New Orleans through the ordinary mail channels. The strife was between several Baltimore papers for the first use of the telegraph between Washington and Baltimore. The Telegraph-office was close to the Post-office, both being more than a mile from the wharf, at which the mail-steamer, after having ascended the Potomac from the Aquia Creek, stopped, and from which the mail-bags had to be carried in a wagon to the Post-office. The plan adopted by the papers to anticipate each other was this. Each had an agent on board the steamer, whose duty it was, as she was ascending the river, to obtain all the information that was new, and put it in a succinct form for transmission by telegraph, the moment it reached Washington. Having done so, he tied the manuscript to a short heavy stick, which he threw ashore as the boat was making the wharf. On shore each paper had two other agents, one a boy mounted on horseback, and the other a man on foot, ready to catch the stick to which the manuscript was attached the moment it reached the ground. As soon as he got hold of it he handed it to the boy on horseback, who immediately set off with it at full gallop for the Telegraph-office. There were frequently five or six thus scrambling for precedence, and as they sometimes all got a good start, the race was a very exciting one. Crowds gathered every evening around the Post-office and Telegraph-office, both to learn the news, and witness the result of the race. The first in, secured the telegraph, and in a quarter of an hour afterwards the news was known at Baltimore, forty miles off, and frequently before the mail was delivered, and it was known even at Washington itself. On an important occasion one of the agents alluded to as being on board, beat his competitors by an expert manœuvre. He managed, unperceived, to take

a bow on board with him, with which, on the arrival
of the boat, he shot his manuscript ashore, attached
to an arrow, long before his rivals could throw the
sticks ashore to which their's was tied.   Next evening, however, when still more important news was
expected, and arrived, he was in turn outwitted.   On
her way up the boat touches at Alexandria, on the
south side of the river, to leave the bags directed to
that town, and take others from it.   On this occasion
one of the newspapers had a relay of horses between
Washington and Alexandria, the rider receiving the
news from the agent on board at the latter place, and
galloping off with it to the capital.   The bow was
then of no use, for by the time the news-laden arrow
was shot ashore, the intelligence designed for the
rival paper was being telegraphed to Baltimore.   It
will thus be seen that the American press partakes of
that " go-aheadism " which characterises the pursuit
of business in so many of its other departments in
America.

A people may very generally be able to read, and
yet the means of intellectual gratification may be
placed beyond their reach.   There can be no doubt
but that it is greatly owing to their cheapness that
American newspapers are so universally perused.
This cheapness arises partly from competition, partly
from the little expense at which newspapers are got
up, and partly from the absence of causes tending
artificially to enhance their price.   But there is no
little misconception in this country as to the cheapness of American newspapers.   The American people
have taken care that no excise or other duties should
exist, which might enhance the price of literature,
in any form in which it might appear.   America

is thus, undoubtedly, the land of cheap literature; but, in connexion with the newspaper press, the mistake made is in supposing that English journals are exceedingly high-priced, as compared with those of America. I shall show that, not only is this not the case, but that independently of stamp and excise duties, the first-class papers of this country are in reality cheaper than the first-class papers in America. It is true that a large proportion of American newspapers are sold at the low rate of two cents, and some at one cent a copy, but it would be unfair to institute anything like a comparison between them and the daily press of this country.

Taking the first-class papers of New York, such as the *Courier* and *Inquirer*, the *Journal of Commerce*, the *Commercial Advertiser*, the *New York American*, &c. we find them sell at six cents per copy. This is about threepence-halfpenny of our money. It is obvious, therefore, that if they had a penny to pay by way of stamp duty upon each number, and about a halfpenny more in the shape of excise duty upon paper, their cost would be *fivepence*, which is the price of our daily papers. So far they appear to be upon an equality. But when we take into account the enormous expense at which a paper in London is conducted; the cost of its parliamentary corps, its staff of editors, and its legion of foreign correspondents; and consider also that, with one exception, the advertising patronage of our daily papers (thanks to the advertisement duty,) is far less than that of the American journals, we see that a London paper with stamp and excise duty off it, and selling at the same price as an American, would, in reality, considering the expensive appliances brought to bear upon it, be much cheaper than the transatlantic journal. But I have not yet done with the points in the comparison favourable to the English press in point of price. Whilst the American papers, had they the same burdens to bear as the English have, would sell at fivepence, the actual selling price of the English paper is *fourpence*. In other words, the selling price, minus the stamp and excise duties, is twopence-halfpenny, or one penny lower than the American paper, which is produced at one-half the expense, so far as all its literary departments are concerned. It is true that to the public the price of a London paper is fivepence, but it is the newsvender, not the newspaper, that pockets the difference. Now the newsvending system in America has made little or no progress, so that a paper selling there at three-pence-halfpenny, enables its proprietors to pocket the whole profits upon the sale, instead of sharing them, as here, with parties intermediate between them and the public. The true state of the case, therefore, between the two papers is this, that whilst a first-class American paper sells for threepence-halfpenny, a London paper which is produced at infinitely greater expense, and has a smaller advertising patronage, and which is at the same time burdened with stamp and excise duties to the extent of nearly a penny-halfpenny per copy, sells at fourpence. Great, therefore, as is the difference of cost in every respect at which they are produced, that in their selling price is but one halfpenny. The difference to the public, but including the newsvender's profit, is three-halfpence.

An English, is thus comparatively cheaper than an American first-class newspaper. It is a pity that by the abolition of those duties which artificially enhance their price, English journals were not nominally as cheap to the public as are American. From making them so, society would reap every advantage. Let it be borne in mind that there is a cheap press in this country, a very cheap press, the issues of which seldom meet the eye of the so-called respectable classes, but which are daily diffusing their intellectual and moral poison amongst the lower orders. And what have we to counteract this great, though but partly appreciated evil? The bane is cheap, the antidote is dear. The bane works, therefore, without check. We cleanse our putrid sewers by directing through them currents of fresh water. Why not bring similar purifying influences to bear upon the daily receptacles of moral filth? We are doing all we can by the erection of baths and wash-houses to superinduce amongst the people a cleanly habit of body, by cheapening the processes by which alone, in the midst of a large community, it is to be attained. But we take no efficient steps to secure for the lower orders a wholesome habit of mind. We make war with physical disease, but leave moral pestilence to do its deadly work. The cheap press, with all its pestiferous influences, is the poor man's intellectual aliment, whilst the respectable and high-priced press is the rich man's luxury. It is essential to the well-being of society that the latter should circulate where the former circulates. It is essential, therefore, to the well-being of society that the respectable press should be made as cheap as possible.

CHAPTER IX.

RELIGION IN THE UNITED STATES.

Separation of the Church from the State.—Effects of this upon both the Church and the State.—Voluntaryism in America.—Difference between Voluntaryism there and Dissent in this country. —Sect in America.—Proportion in point of numbers and influence of the different Protestant Sects.—The Roman Catholics. —Far-seeing Policy of the Church of Rome in America.—Revivals.—Independence of the American Clergy.—Zeal of some of the American Churches.—Attention paid to Strangers in American Churches.—Church Music.—The Organ.—Sunday Schools.—Conclusion.

WHILST education is universally promoted in America by the State, as a matter in which the State is equally interested with the individual, religion is left to itself, not as a matter in which the State has no interest, but as being of such high individual concern, that it is thought better for the State to keep aloof and leave it to the care of the individual. Moreover, the experience of other nations had taught the Americans, ere they framed their constitution, that religion and politics were not the most compatible of elements, and that political systems had the best chance of working smoothly towards their object, when least encumbered by alliances with the church. If there was one thing on which, more than another, they were agreed, in preparing a political frame-work for the Union, it was in the propriety and necessity, if they would not mar their own work, of divorcing the State from the Church. The ceremony of separation may be delayed in countries in which the connexion exists, long after the necessity for it is felt and its propriety acknowledged, from the difficulty which is ever in the way of breaking up old ties and associations. It is thus that the alliance between Church and State in England is likely, for some time, to outlive opinion in its favour. Were we forming our political system anew, there is no doubt but that many who are now Church and State men from circumstances, would be anti-Church and State men on principle. The connexion in England now depends for its continuance more upon the conservative feeling which instinctively rallies round an existing institution, no matter how unnecessary soever it may be, or how ill-adapted to the circumstances of the time, than upon any very prevalent conviction of its being beneficial to religion, or advantageous to the State. The Americans were fortunate in determining and arranging their system, in having a clear field before them. In settling it, they were at liberty to base it upon their convictions untrammelled by inconvenient pre-existing arrangements. They, therefore, wisely determined to leave out of their plan, a feature, which, as it seemed to them, had added neither strength nor harmony to the political systems of others. They not only divorced the State from the Church, in a strictly political sense, but, in so doing, refused to allow the Church a separate maintenance. Her empire they regarded as the heart of man, and if she could not establish herself there, they would not sustain her in a false position. Thus, whilst we bolster up the Church and leave education to take care of itself, they promote education as a people, and leave religion to its own elevated sway over the individual. Thus far they come in aid of Christianity—that in educating the people, they prepare the public mind for the more ready reception of its lofty inculcations and sublime truths. But they go no further. The social duties which man owes to man, the State will enforce. But if the people forget their duties to God and to themselves, it is God that must deal with them, and not the State. To make religion in any degree a matter of treaties, protocols, and statutes, is to detract from its high moral dignity—to make it a matter of State convenience, is to abase it.

There is no principle more freely admitted, both practically and theoretically, in America, than the right of every man to think for himself on all matters connected with religion. The side from which they view the matter, is not that the admission of this principle is a concession made by each to all, and by all to each, but that the denial of it would be an indefensible invasion of one of the highest rights of the individual, a right superior and antecedent to all social and political arrangements. It is thus that the insulting term "toleration" is but seldom heard in America in connexion with the religious system of the country. To say that one tolerates another's creed, implies some right to disallow it, a right that happens to be suspended or in abeyance for the time being. The only mode in which the Americans manifest any intolerance in reference to religion is, that they will not tolerate that the independence of the individual should, in any degree, be called in question in con-

nexion with it. They will therefore tolerate no
political disabilities whatever to attach to a man on
account of his religious belief. In their individual
capacity, they seek not to coerce each other's opinions;
in their social and political capacity, they regard each
other as citizens, and simply as such. If a man per-
forms the duties, and bears the burdens of a citizen,
they do not inquire into his views upon the Trinity,
or his notions of the Immaculate Conception.

This state of things will give rise, in the reader's
mind, to two questions—What does the State gain by
it, and what is its effect upon the interests of religion? The State must be a gainer by the removal
of a prolific cause of discord and bitterness of feeling.
In addition to this, the Union has the exalted satis-
faction of knowing that it has washed its hands, in
its political capacity, of every thing savouring of
religious persecution, whilst the Americans, as a
people, are not liable to the scandal of seeking per
force to save one man through the medium of another's
views. Its effect upon the interests of religion will
be seen from what follows.

Well may the nations of the world fix their eyes
anxiously on America, for it is the scene, not only of
a great political, but also of a great religious experiment. The problem which it is working out involves
political liberty in connexion with society, and the
voluntary principle in connexion with religion. For
the first time since its junction with the State, has
Christianity been thrown upon its own imperishable
resources, in the midst of a great people. And has
it suffered from its novel position? Who accuses
the Americans of being an irreligious people? Nay,
rather, who can deny to them, as a people, a pre-

eminence in religious fervour and devotion ? There are many who regard religion as very much a matter of climate, and believe that it is more likely to find a welcome in the reflective minds and comparatively gloomy imaginations of the inhabitants of the North, than in the quicker wits and more lively fancies of the denizens of the South. Whatever be its cause, the further north we go in our own country, the more do we find the people imbued with the religious sentiment, and the more universally do we find them submitting to the dominion of religion. It is precisely so in the United States. The North, as it is the more energetic, is also the more religious section of the Confederacy, there being as great a difference in connexion with religion between the New Yorker and the Carolinian, as between the rigid and morose Presbyterian of Glasgow and the more cheerful Churchman of London. To whatever extent religion may have laid hold of the public mind, in this or in that section of it, the voluntary principle is ubiquitous throughout the Union. If in the North the obligations of religion are extensively, so are they voluntarily observed ; if in the South they are comparatively neglected, they are voluntarily overlooked. There is no State Church in the one case to take credit for men's zeal, and in the other, to receive blame for their callousness. The same difference is observable in both countries in connexion with latitude. But taking each country as a whole, the religious sentiment is more extensively diffused, and more active in its operations in America than in Great Britain. And this, in a country in which religion has been left to itself.

What then becomes of the sinister predictions of those who assert that a State connexion is necessary to the vigorous maintenance of Christianity? Does religion assume a languid aspect in America, where there is no such connexion? Is it less vigorous in Scotland than in England, the alliance in the former being but partial as compared with its closeness and intimacy in the latter? Throughout New England, the northern, and some of the middle States, religion is not only as active, but it is as well sustained as it is in this country, notwithstanding the aid and comfort which it here receives from the State.

Are proofs of the vitality and energy of religion in America wanted? Look at the number of its churches, the extent and character of its congregations, the frequency of its religious assemblages, the fervour of its religious exercises, and the devotion of its religious community, testified by their large and multifarious donations for religious purposes both at home and abroad. Like the Church in Scotland, the Church in America too has its great schemes, towards the maintenance of which it is constantly and liberally contributing. It has its Missions, home and foreign, its Bible and Tract Societies, its Sunday School Unions, and associations for the conversion of the Jews; in short, there is not a scheme which has of late interested the Christian world, in which it does not take a cheerful and prominent part. Does this bear out the assertions of those who say that the voluntary system has a paralysing influence? But we need not go to America for a practical refutation of this oft asserted fallacy. It is amply furnished to us at home, for by far the most energetic section of our Christian community is that which constitutes the great voluntary body. The proofs are all in favour of the converse of the proposition; everything, both here and in America, tending to show that the religious sentiment is more diffused and energetic, when allied to voluntaryism, than when it is taken under the protection of the State.

It is only in America, however, that the voluntary principle has had an opportunity of exhibiting itself in its proper character. There are many, judging of it from the phase which it assumes in this country, who object to it, on the ground of its apparent tendency to run into fanaticism, and to carry that fanaticism into politics. In a country divided between the voluntary principle and that of an established Church, the tendency to over-zeal and fanaticism is much increased, by the conflict which is waged between the two principles. The blood of the attacking party is always more heated than that of the attacked. The voluntaries here are the attacking party. The Church, with some slight exceptions, remains on the defensive, the cohorts of voluntaryism assailing her at every practicable point. Their favourite tactics consist in outstripping her in zeal and devotedness—no very difficult matter; but zeal once roused, and inflamed by resistance, frequently runs into extremes, which it never contemplated in its cooler moments. Thus the voluntary churches, in running a race in zeal with the Church, get into such a habit of racing that they throw down the gage to each other. Zeal thus rises into enthusiasm, and enthusiasm often merges into fanaticism. No matter from what point it starts, when religion reaches this point it becomes bigoted, relentless, intolerant, and persecuting. It also transcends the line of its own duties, and whilst repudiating all connexion with the State, would fain reduce the State into subjection to it.

Forgetful of its own vocation, it intermeddles with matters of a purely secular character, and thus, instead of aiding men in their career of social advancement, frequently throws the greatest stumbling-blocks in their way. It is thus that religion, in both its established and voluntary phases in this country, has proved itself the greatest drawback to education. Churchmen and Voluntaries seeking to make it exclusively subservient to their own views, instead of renouncing all connexion with it as religionists, and treating it as primarily a matter of secular concern.

Voluntaryism in America exhibits itself in a more attractive aspect. There it has the whole field to itself, and its manifestation of a more tractable disposition is owing not a little, perhaps, to the absence of those inducements to strife and opposition to which Dissent in this country is exposed. Let me not be here understood to mean that religion, in the different forms in which it manifests itself in America, is always characterised by that gentle, placid and forbearing spirit which it should ever seek to cherish. It is frequently as much inflamed by zeal and distorted by fanaticism as it is here; but there are directions in which a misguided zeal often tends in this country which it never takes in America. Here it frequently applies itself to political objects, there it scarcely ever does so. An American zealot may be quite reasonable as a politician, because, in his capacity of zealot, he seldom encounters a political opponent. Sects in America contend with each other almost exclusively on the religious arena, their great object being to outstrip each other in fervour and devotion, partly from the desire to spread what is sincerely believed to be the truth, partly from the pride which

mingles with belief, and partly from the desire to
increase the number and social influence of the sect
throughout the Union. Religion in America is
rarely brought into the field as a political accessary.
Americans seek not to achieve anything political
through its means. In this respect, religion escapes
in America the degradation to which it is so frequently subjected here. By refraining from interfering with politics, and confining itself to a purely
social influence, it recommends itself more to the community generally, than it would do were it, as in this
country, constantly thwarting the progress of secular
interests. So little is it the habit of voluntaryism in
America to interfere in matters of a political bearing,
that when Congress, although the great majority of
the members were Protestants, selected a Roman
Catholic priest as one of its chaplains, no one dreamt
of organizing a religious agitation to prevent such an
infraction of Protestant privileges in future. Thus
both the State and the Church find it to their account
to confine themselves exclusively to their respective
provinces, the State abstaining from all interference
with religion as the State, and the Church taking no
part as the Church in the management of secular
affairs. Voluntaryism in America is, for this reason,
divested of many of those features which render Dissent unattractive to such numbers in this country.
It is when in forced or accidental connexion with
politics that sect exhibits itself in its most repulsive
aspect. Where one denomination has a political
side, others have, by consequence, the same. They
mutually assail each other, the one to maintain its
privileges and extend its power ; the others, to defend themselves against coercion, and deprive their

rival of its usurped authority. The strife not being of an exclusively religious character, the passions of men are not kept in that check which decency enjoins upon them in a purely religious contest. They thus learn to carry into questions of a direct religious bearing, if not exclusively of a religious character, all the excitements and passions of political contention. Such is Dissent in this country—circumstances have made it so. Voluntaryism in America, being subjected to fewer causes of disturbance, is more placid in its action, and more engaging in its demeanour. Sect there, as here, is in constant rivalry with sect, but the race they run with each other being chiefly a religious one, their conduct in pursuing it is more consistent with their professions, and more in harmony with the spirit of genuine religion, than is the case in the warfare waged against each other by denominations in this country. In America, the only disturbing influences to which sect is exposed, are religious zeal and fanaticism; whereas in this country, when these are dead, it is frequently roused into phrenzy by political excitements. Let voluntaryism therefore not be judged of solely from its manifestations in this country, where there are so many influences at work inimicable to its more favourable development.

The reader has scarcely to be told that nine-tenths of the American people are Protestants. The number of sects into which they are divided and subdivided can only be ascertained by a patient investigation of the census. There is no country in the world in which sect flourishes more luxuriantly than in America. This is perhaps the natural result of that freedom of opinion on matters of religion, which is one of the chief characteristics of the Protestant mind. Nor does any harm accrue from it, when sects are not brought into collision from causes with which religion has, or should have nothing to do, seeing that the points on which they differ are in reality, in nine cases out of ten, of comparatively minor importance. They suffice nevertheless to separate sect from sect, and to engender between them that spirit of rivalry which some regard as advantageous to the spread and preservation of the truth, each sect keeping a vigilant and jealous eye on the creed and inculcations of its rivals.

As regards numbers, the following is the order in which the principal sects range in America. First come the Methodists, who have upwards of 7,000 ministers, and more than 1,200,000 communicants. Next in order are the Baptists, divided, like the Methodists, into numerous sub-sects, and having about the same number of ministers, but not quite so many communicants. After these come the Presbyterians, divided into the New school and Old school party, the quarrel between them having partly arisen from slight doctrinal differences, and partly in connexion with some property. The former is the more numerous section, having about 1,700 ministers, and nearly 200,000 communicants; the latter, about 1,300 ministers, and 150,000 communicants. United, they have about 3,000 ministers, and 350,000 communicants. The Congregationalists follow next in order, having about 1,800 ministers, and a little upwards of 200,000 communicants. These are subdivided into the Orthodox and Unitarian Congregationalists, the latter having nearly 275 ministers, and 40,000 communicants. The Evangelical Lutherans follow next in order, a denomination chiefly composed of German emigrants and their descendants. They have about 500 ministers, and 145,000 communicants. The Episcopalians follow, with upwards of 1,300 ministers, and about 80,000 communicants; immediately after whom come the Universalists, with more than 700 ministers, and upwards of 60,000 communicants. It is needless to trace the relative standing of the minor sects. New England is the chief seat of Congregationalism in its two phases, Orthodox and Unitarian, and of Universalism. The principal stronghold of Presbyterianism is in the Northern States, although in no other part of the Union does any one denomination so completely predominate as the Congregationalists do in New England. The wealth, fashion, and intelligence of that part of the country are included within this denomination, although, taking the country generally, the predominance of wealth and intelligence is with the Presbyterians, notwithstanding that they rank but third in point of numbers. The Episcopalians are comparatively few in number, but there is much wealth and intelligence with them. With the exception of the judges of the Supreme court at Washington, I never beheld a civil functionary in America decorated with any of the paraphernalia of office to which the European eye is so accustomed. With the exception of the Episcopal clergy in America, I never saw a Protestant minister wear either gown or surplice in the pulpit.

In the above enumeration the Roman Catholics have not been mentioned, confined as it has been to the Protestant sects. Their numbers are not great, as compared with the Protestants; but they are nevertheless a sect of considerable power in the Union. In 1848 they had about 850 churches—nearly 900 priests, and 1,175,000 communicants. It would not be correct, however, in comparing their aggregate number with that of the other sects, to take the number of communicants as the basis of comparison; inasmuch as with the Roman Catholics almost every adult is reckoned a communicant, which is far from being the case with the adherents of the Protestant denominations. The Catholics are a strong body in all the large towns; and in some parts of the country they have rural districts, of considerable extent, under their sway. Until the purchase of Louisiana, in 1803, Maryland was, in point of numbers, the leading Catholic State in the Union, as she is yet in point of influence. As the American has not yet out-numbered the French population in Louisiana, it follows that the largest moiety of the white inhabitants of that State are Roman Catholics.

It is to her colonial origin that Maryland owes the pre-eminence which she has so long maintained as the chief seat of Roman Catholic influence in the Union. After other sects had fled from the Old World to the New to escape persecution, the Catholics, in some instances, found that they too were in want of a place in which they could worship God according to their consciences. They accordingly emigrated in great numbers to the State of Maryland, named after Queen Mary, and being for some time a proprietary colony belonging to Lord Baltimore, whose name its chief town still bears. The Roman Catholic colonists set an early example of religious toleration, which was but ill requited by the Protestants, as soon as they attained a numerical superiority in the State. The number of Roman Catholics in the State is now daily diminishing, as compared with that of the

Protestants—the hold which Catholicism now has of Maryland consisting chiefly of the adhesion to it of many of the older families of the State. The Catholic cathedral at Baltimore has already been adverted to, in the brief description given of that city. The only other ecclesiastical edifice in the Union, dedicated to Catholicism, which deserves the name, is the cathedral at New Orleans.

It is not so much on account of its present number of adherents, or the influence which it now exerts, that Catholicism in the United States demands the attention of Christendom. It is in view of its future prospects, that it assumes an attitude of rather a formidable character. Nowhere on earth is the far-seeing policy of the Church of Rome at present so adroitly displayed as on the American continent. Indeed from the earliest epoch of colonization we find her aiming at the religious subjugation of America. For a time success seemed to crown her efforts. The whole of South America, Central America, and the greater part of North America, together with all the islands on the coast, were divided between the crowns of Portugal, France, and Spain. England, for many years after her first attempts at colonization, possessed but a comparatively narrow strip of land between the Atlantic and the Alleganies, and extending along the sea-board from Acadia to Georgia. New France swept round the English colonies, from the mouth of the St. Lawrence to that of the Mississippi, whilst the Spanish Floridas intervened between them and the Gulf of Mexico. Within this wide embrace, with the ocean in front, lay the group of Protestant colonies belonging to England. It was not sufficient for the Church of Rome that she hemmed them in on three sides by her territory. The wide domain which owned her sway was but thinly peopled, whilst the English colonies were rapidly filling with population. Protestantism was thus fast attaining on the continent a more extensive moral influence than its competitor. It was then that a Roman Catholic colony was planted in its very midst, on the shores of the Chesapeake, the policy of the church having had no little influence on the moral destinies of Maryland. But the tide had set in too strongly in favour of the rival system, and it soon overpowered all opposition to it. Since that time Catholicism in Maryland has acted more on the defensive than otherwise—its object having chiefly been to maintain itself as a centre and rallying point for Catholicism in the Union, with a view to future operations in new and vaster scenes of action.

The ground has now for many years been broken, and these operations have long since actively commenced. The Roman Catholic church has, in a manner, abandoned the comparatively popular States of the sea-board, and fixed its attention upon the valley of the Mississippi. In this it has discovered a far-seeing policy. Nineteen-twentieths of the Mississippi valley are yet under the dominion of the wilderness. But no portion of the country is being so rapidly filled with population. In fifty years its inhabitants will, in number, be more than double those of the Atlantic States. The Church of Rome has virtually left the latter to the tender mercies of contending Protestant sects, and is fast taking possession of the great valley. There, opinion is not yet so strongly arrayed against her, and she has room to hope for ultimate ascendency. In her operations, she does not confine herself to the more populous portions of the valley, her devoted missionaries, penetrating its remotest regions, wherever a white man or an Indian is to be found. Wherever the Protestant missionary goes he finds that he has been forestalled by his more active rival, whose coadjutors roam on their proselytizing mission over vast tracts of country, into which the Protestant has not yet followed him with a similar object. Catholicism is thus, by its advance-guards, who keep pace with population whithersoever it spreads, sowing broad-cast the seeds of future influence. In many districts, the settler finds no religious counsellor within reach but the faithful missionary of Rome, who has thus the field to himself —a field which he frequently cultivates with success. In addition to this, seminaries in connexion with the church are being founded, not only in places which are now well filled with people, but in spots which careful observation has satisfied its agents will yet most teem with population. Ecclesiastical establishments too are being erected, which commend themselves to the people of the districts in which they are found by the mode in which they minister to their comforts and their necessities when other means of ministering to them are wanted. The Sisters of Charity have already their establishments amid the deep recesses of the forest, prescribing to the diseased in body, and administering consolation to the troubled in spirit, long before the doctor or the minister makes his appearance in the settlement. By this attention to the physical as well as to the moral wants, the Roman emissaries, ere there are yet any to compete with them, gain the good will of the neighbourhood in the midst of which they labour, and proselytism

frequently follows hard upon a lively sentiment of
gratitude. Circumstances have favoured the Church
of Rome in the development of this policy. When
both the St. Lawrence and Mississippi, with most of
their tributaries, were in the possession of France, a
belt of ecclesiastical establishments accompanied the
chain of military posts, which, extending westward
from the coast of Labrador to the lakes, descended
thence to the mouth of the Ohio, and then spread
north and south on both banks of the Mississippi.
The basis was then laid for the future operations of
the Church. It is nearly a century since France lost
Canada, since which time a gap intervened between
the Church's establishments in its eastern section and
those dotting the province of Louisiana. But down
to the year 1803, the whole of the west bank of the
Mississippi, and both banks in the neighbourhood of
its mouth, were in the hands of the French, the advanced posts of the Church spreading and multiplying
between St. Louis and New Orleans, whilst the
eastern or Protestant bank of the river was yet an
unbroken wilderness. The present operations of the
Church of Rome, therefore, in the valley, cannot be
regarded as an invasion of that region, her object
now being to profit by the advantages which she so
early secured. Were the Protestant sects to confront
her as actively as they might, in the great field which
she has thus selected for herself, they might even yet
check her growth and limit her influence. But they
seem to be either unaware of, or indifferent to, the
danger with which they are menaced. They are
seeking to rival each other in the older States, whilst
their common rival is laying a broad foundation for
future influence in that region, which will soon

eclipse the older States, in population at least. Both in St. Louis and New Orleans, some of the best seminaries for young ladies are Catholic institutions, and not a few of those who attend them become converts to the Church. But it is in the remote and yet comparatively unpeopled districts that the probabilities of her success in this respect are greatest. She has thus, in the true spirit of worldly wisdom, left Protestantism to exhaust its energies amongst the more populous communities; and going in advance of it into the wilderness, is fast overspreading that wilderness with a net-work which will yet embrace multitudes of its future population. How can it be otherwise when, as settlements arise, they find at innumerable points the Church of Rome the only spiritual edifice in their midst. Were she to secure the valley, she would gain more in America than all she has lost in Europe. The stake is worth striving for; and Protestantism would far more consult its own interests by directing its efforts less to the Niger and more to the Mississippi.

For a long time a strong aversion to the Americans actuated the French settlements, an aversion chiefly founded upon religious considerations. The priesthood regarded republicanism as inimical to the hierarchy, and imbued their flocks with the same belief. The existence of the belief that a connexion with them, if successful, would be inimical to the interests of the Church, was one of the chief sources of the loyalty which the French Canadians exhibited in refusing to join the revolutionary movement on which the Protestant colonies embarked in 1776. The purchase of Louisiana, however, and its incorporation for the last forty-five years with the Union, have greatly tended to weaken this belief, and to eradicate from the Catholic mind in America the aversion which it once entertained to a political connexion with the Republic. The *habitans* are now rapidly reconciling themselves to the idea of such a connexion. The same feeling actuated, and to some extent still actuates, the Mexicans. Apprehensive that the war which broke out in 1846 might end in the entire subjugation of their country, the Mexican hierarchy sent emissaries into the Union to ascertain the precise effect which such an event would have upon the Church; and, from all I could learn at the time, they returned with their fears, if not wholly removed, at least greatly diminished.

I have already intimated that sect in America is not wanting in occasional ebullitions of zeal and fanaticism. Indeed with some sects fanaticism sometimes attains an extravagance which borders on the sublime. As violent fits could not last long without exhausting the body, so these periodic religious spasms—fortunately for the sanity of the public mind,—although they pretty frequently occur, are only temporary in their duration. Some sects are cooler in their moral temperament than others, and are seldom or ever affected by them; but others are afflicted with them almost with the regularity, though with longer intervals between, of shivering fits during an attack of ague. The denominations most unfortunate in this respect are the Baptists and Methodists, whilst occasionally the more sober Presbyterians sympathise and fall a prey to the disorder. The moral distemper which on these occasions seizes upon masses of the population, is termed a " revival." Such visitations are not rare amongst ourselves, but it is seldom that they attain anything like the appalling influence which they sometimes gain in America. Like a physical epidemic, their course is uncertain and capricious, frequently attacking communities which have always been ranked amongst those morally healthy, and passing over, in reaching them, others which had previously exhibited themselves in a state of almost chronic religious derangement. These revivals, when they occur, at first generally embrace but one sect; but if they take hold of the public, they soon draw other denominations into the movement, which do not, however, throw aside their distinctiveness in taking part in it. The interest of the whole affair is almost invariably centered in one peripatetic enthusiast, who, watching the tone and temper of the public mind, takes advantage of the existence of a pervading *ennui,* and commences a religious campaign when any novelty is sure to recommend itself. For a while his success may not appear to be commensurate with his efforts, but by-and-by the locality in which he labours is roused, the movement spreads into the adjoining districts, the revival acquires momentum, and millions are in a frenzy. The enthusiast proceeds on his tour of moral disturbance and religious agitation. In each locality which he visits, the nucleus of the movement is the sect to which he belongs. Most of the day is divided between prayer-meetings and sermons. People get nervous, and the malady spreads. Members of other denominations flock to the church, some from curiosity, others from different motives. The lion of the movement is in the pulpit, sometimes foaming at the mouth in the midst of his declamations. The weaker members of the dense congregation yield—they get agitated and alarmed—hysterics follow, and some are in tears. The sympathy of numbers tells upon many more—hopes are inspired and alarm engendered, which bring them back again to the scene, to be similarly influenced as before—and they end by seeking the " anxious seat," confessing their sins, and being " born again." Business is neglected, families are divided and disturbed, and the greater part of the community, but not until their nerves are almost shattered, give way more or less to the reigning fanaticism of the hour. Hundreds are added to the Church in a day.

During one of these revivals, which it was my lot to witness, and of which the Baptist denomination was the *primum mobile*, I knew as many as five hundred baptized in the course of three hours, in a huge tub which was kept at the foot of the pulpit for the purpose. Rakes are reclaimed, prodigal sons return to their long-neglected duties, backsliders make open confession of their sins in the church and are reinstated, and hundreds who have been hitherto indifferent give way to the fervour of the hour. And this is what is called making converts. The consequences almost invariably prove how great a mistake is made in this respect. Numbers of those who, yielding to an impulse engendered more by a physical excitement than anything else, in the moment of dread or enthusiasm, enrol themselves as converts, relapse into their former ways as soon as the paroxysm is passed, and the reaction ensues. They do worse than this, for a backward step taken under such circumstances is tantamount to several under ciscumstances of an ordinary description. There is then the pernicious example which they set to be taken into account, and the readiness with which the scoffer seizes upon their backslidings to throw ridicule upon religion itself. When will these zealots learn that religion is a matter of the judgment as well as of the feelings! Yet the whole of their system of revivals is built upon an exclusive appeal to the weaker side of man's nature. They may trample upon the judgment for the time being, but they cannot always keep it in thrall; and when it does assert its supremacy, it may avenge itself for having been dragged into one extreme by permitting itself to be hurried into another. The principle of these audacious caricatures upon religion is not, " come, let us reason together," but " come, and be scared into conversion." The fanaticism which they engender is fierce whilst it lasts; but the reaction, which is not long delayed, does incredible mischief to the cause of rational religion.

The most enthusiastic revival ever witnessed by me had its inception amongst the Baptists. It commenced somewhere in the West, and spread in an incredibly short space of time over a large portion of the Northern States, embracing at last the adherents of almost every sect within its influence. The source of this moral perturbation was an Elder belonging to the denomination named, who made the tour of the North and North-west. Wherever he went, he soon managed to engender a perfect *furore*, thousands flocking to hear him rave, and hundreds being almost daily frightened by him into repentance and regeneration. A large proportion of the residents of each town in which he pitched his tent for a time were excessively annoyed, inconvenienced, and scandalized by the proceedings which accompanied his sojourn, and one had cause to be thankful in walking the streets if he escaped impertinent encounters by the way. I was myself frequently stopped on the public pavement by parties whom I knew not, and admonished to repent, and go and be baptized. On one occasion I was met and accosted by the Elder himself.

" Young man," said he, stopping me, and laying his hand paternally upon my shoulder, " how's your soul ? "

" Quite well, I thank you," I replied,—" how's yours ? "

" Bless the Lord ! " he continued.

" Amen ! " I responded.

" You're an heir of damnation," said he in great haste, after apparently measuring me from top to toe with his eye.

" The idea seems to give you positive pleasure," observed I.

He looked at me again for a few moments, after which he told me in great confidence that the sons of Anak would be brought low. To this I replied that, not knowing them, I could not be expected to feel much interest in their fate.

He looked hard at me again for a few seconds, and then shouted so as to attract the attention of the passers-by—" You're a Scribe—you're a Scribe ! "

" Anything but a Pharisee," I replied, and walked on, leaving him to make what application he pleased of my response.

He was very successful in his agitation whithersoever he went, throwing town after town into paroxysms of excitement, and securing in each a great many converts for the nonce. The per-centage of them who shortly afterwards became backsliders was very great. It seemed to be his peculiar delight to vulgarize religion as much as he could, frequently making use of similes which bordered on ribaldry, and sometimes even on blasphemy. On one occasion, being tired of the gospel, he betook himself to slander, telling his hearers in one breath to be forbearing and to love one another, and in the next indulging in the most uncharitable suspicions of his neighbours. Amongst others whom he slandered was an hotel-keeper, who also became the victim of the malicious innuendos of his chief disciple. This gave rise to two parties in the community, the enthusiasts rallying round the Elder, and the " ungodly," as they were termed, ranging themselves under the standard of the injured party. The more orderly and decorous portion of the inhabitants kept themselves aloof from both parties. At length the time of the Elder's departure drew near, and it was known that his chief disciple was to accompany him. A disturbance of the public peace was apprehended, and the friends of order advised them to depart secretly. This they refused to do, persisting in their resolution to go at the time fixed upon by the regular stage. The morning of their departure was one of commotion bordering on riot. The " ungodly " had procured a wagon, which they filled with musicians, who rode up and down the street in which the obnoxious individuals were lodging, playing the Rogue's March. It was not until they had both got into the stage and were about to depart, that the disciple was arrested in an action of slander, at the suit of the aggrieved inn-keeper. Both he and the Elder, as well as their numerous abettors, gloried in this; it was persecution, and of itself testified to the high origin of their mission. Bail was soon procured, and the parties permitted to proceed on their way, the musicians following them out of the town with no very complimentary airs. Some months afterwards the action came on for trial in the same place. The Elder was the chief witness on the part of the defendant. When in the witness box, he was asked by the counsel for the plaintiff, if he had not had reason to believe that his departure, unless private, would occasion some display inimical to the public peace? He said he had been informed to that effect.

"Were you not advised to depart secretly?" he was asked.

"I was," replied he.

"And why did you not do so?" was the next query put to him.

"Because I was determined to have my way," he replied, "and to let the devil have his."

In commenting upon this part of the evidence, the counsel for the defendant emphatically approved of the Elder's determination to make an open and public exit from the town, even at the risk of a disturbance of the peace, citing the conduct of Nehemiah in his justification, who, when advised to fly from the enemies of the Lord, refused to do so. But the opposite counsel was not to be put down by such authority as this, and contended that if scriptural precedent was to be relied upon, it must follow the rule of precedents in other cases, which is that, *ceteris paribus*, the latest shall rule. The case of St. Paul, he maintained, was more binding because more recent than that of Nehemiah, the great Apostle having been let down from the walls of Damascus in a basket, when his exit otherwise from the city might have involved a violation of public order. In the sight of the audience this gave the whole matter rather a ludicrous turn, judges, jury, bar, and spectators smiling at the retort. It was received by the community in the same spirit, and treated as a good joke, and did much towards undoing the effects of the Elder's preaching. It is not always that revivals lead to such scenes, but they are generally accompanied by a degree of fanaticism and intolerance truly deplorable. They disturb the peace of families and unsettle the ordinary relations of society. Happily their effects are evanescent, or they would be the more to be regretted. Nor are they always so violent as some that I have seen. Occasionally they are what would be denominated failures, from being attempted when the public mind is not in proper tune for them. The most decorous are those which originate with the Presbyterians.

There are many in this country who fall into the mistake of supposing that the voluntary system, as developed in America, is utterly incompatible with that degree of independence on the part of the clergy, which is necessary to enable them efficiently to perform their duties. Amongst others who have fallen into this error is Lord John Russell, who, notwithstanding his vast and varied information on other subjects, is generally at fault when he undertakes to speak of the United States. I have heard him in the House of Commons illustrate his argument that voluntaryism was inconsistent with clerical independence, by alluding to the condition of the clergy in America, who, he contended, were so utterly dependent for subsistence upon their flocks, that they dared not reprove them in the manner in which a pastor should sometimes deal with his people. If their language in the pulpit, and their conduct in the performance of what may be designated as the more private duties of the clergyman, are to be taken as affording an indication of their independence or subserviency, it would not be easy to find a bolder or less scrupulous set of preachers than those who fill the American pulpits. So far from dealing leniently with the shortcomings of their congregations, they deal with them in a manner which many Englishmen would regard as decidedly offensive. Whatever may be the vices of voluntaryism in America, it cannot properly be alleged against it that it muzzles the clergy.

I have already alluded to the number of religious and benevolent schemes to which the various churches in America very liberally contribute, as evidence of their zeal. Farther proof of this is found in the frequency with which, in some instances, they give themselves up to their religious duties. I have several times heard announcements to the following effect made from the pulpit on Sunday: "On Monday evening the usual monthly prayer meeting in behalf of foreign missions will be held, when a subscription will be taken in aid of the missions. On Tuesday, the Maternal Association will be held at Mrs. So-and-so's. On Wednesday, the usual weekly service will take place in the school-house adjoining the church. On Thursday, the Dorcas Society will meet at Mrs. ——'s. On Friday, will be held the ordinary Sunday-school teachers' meeting; and, on Saturday, district prayer meetings will take place at——" (here would follow a number of places in different districts). And all this in addition to three services on Sunday, and a Sunday-school also to attend to. It always appeared to me, on these announcements being made, absorbing as they did every evening in the week, that the fourth commandment ran great risk of being violated in its second clause, "six days shalt thou labour and do *all* thy work."

American churches are in general neatly built, and look very light and airy. In summer it is absolutely essential that they should be well ventilated, as the heat is often oppressive. There is scarcely a pew but adds to its other appendages one or more large feather fans, and the effect of seeing them all waving at once, from the commencement to the end of the service, is at first both striking and curious. After using it for a few minutes one passes the fan to his or her neighbour. In winter, again, the churches are, in the north, well heated with stoves, in addition to which many families bring with them small tin stoves containing charcoal embers, with which they keep their feet warm, passing them from one to the other as they may be required. The pulpits are quite a contrast to the confined boxes, generally looking like casks, from which clergymen in this country almost invariably address their hearers. The American pulpit is more like the bench in a court of justice, being almost always open at both sides, and being sufficiently spacious to contain six or eight clergymen at a time. In most of the Presbyterian churches the congregations face the doors, so that a stranger on entering finds himself confronting the whole audience. This is at first rather awkward, but it serves this good purpose, that the regular sitters see him at once, and are ready on all sides to offer him a seat. The attention thus paid to the stranger in church is almost universal in America. Frequently have I seen a whole family leave their own pew, and scatter themselves amongst their friends, in order to accommodate a number of strangers entering together and forming one party.

Nor is the organ in America confined to the Episcopal or Catholic churches; it is to be found in the Presbyterian, the Baptist, and the Methodist church, whenever the means of the congregation enable them to have it. Their having it, or not having it, is not a matter of principle, but merely a question of expense. This of itself would suffice to account for the superiority of the music in the American churches to that in the dissenting churches here. But in addition to the possession of an organ, almost every church has its choir, which is not composed of hired musicians, but generally consists of the most respectable members of the congregation, male and female, capable of singing well. By introducing the organ, the Americans very properly avail themselves of a great aid to devotion, in doing which they set a lesson both of prudence and good sense to their self-righteous brethren in this country, who are magniloquent in the confession of their moral weaknesses, but who, at the same time, repudiate everything which might tend to strengthen them. The poor rebellious heart of man frequently requires something to solemnize and attune it for devotional exercises, and this he finds in the awe-inspiring aspect of the cathedral, and in the deep tones of the organ reverberating through the aisles. But some of our modern Pharisees would counsel us to reject as spurious the devotional feelings originating from such sources, and to trust like them to our own righteousness and to the strength of our own purposes. How far this may be presumption, and the other course the want of vital religion, let each judge for himself.

Notwithstanding the rivalry existing between sects in America, they frequently manage to suppress it to some extent so far as their teaching of the young is concerned. I have already shown how far secular education has been divorced from sect, and rescued from its obstructive influences. It is in the Sunday-schools that the youthful mind is imbued with the dogmas of sect, each denomination contenting itself, so far as the education of youth is concerned, with the influence which they obtain over them in these schools. But the Protestant sects frequently unite in Sunday-school demonstrations, when the children from all the schools are collected together under their teachers, and examined and addressed by the clergy of the different denominations. I have sometimes seen them marching in thousands to the largest place of worship for this purpose. There they were, embryo Christians, it was to be hoped, but certainly the germs of future Baptists, Presbyterians, Methodists, New Lights and Old Lights, Congregationalists, Lutherans, &c.

In conclusion, let me remind the reader that, notwithstanding the hot race of competition which it sometimes runs, and the social and individual tyranny of which it is occasionally guilty, sect in America is not the embittered and envenomed thing that it is in this country. If voluntaryism has vices inseparable from its very nature, they are not aggravated there, as here, by extraneous causes already explained. It may be over-zealous, fanatical, jealous, and sometimes even malignant, in its manifestations ; but its evolutions are chiefly confined in America to the religious arena, it being extremely seldom that it is found stepping aside from its own sphere to jumble religion and politics together, and to aggravate the *odium theologicum* by adding to it the acerbities of political contention.

CHAPTER X.

LOWELL.—MANUFACTURES AND MANUFACTURING INTERESTS OF THE UNITED STATES.

Journey from New Haven to Worcester and Boston.—Proceed to Lowell.—Appearance of Lowell.— Its rapid Growth.—Colonial Manufactures.—Difficulties with which they had to contend.—Progress of American Manufactures during the War of Independence, and that of 1812.—Motive Power used in Lowell, and means of employing it.—The Operatives of Lowell.—Educational and other Institutions.—The different Manufacturing Districts of the Union. —New England.—The Northern Atlantic States.—The Southern Atlantic States.—The States on the Mississippi.—Distribution of Manufacturing Capital throughout the Union.—Rise of Cotton Manufacture in America.—Exports of Cotton Goods.—Progress of other Manufactures. — Steam *v.* Water Power. — Comparative strength of the Manufacturing and Agricultural Interests.—The dream of Self-dependence.—The Future.

FROM New Haven I proceeded through the interior of Connecticut to Worcester, in Massachussetts, and thence direct, by railway, to Boston. Almost every inch of this portion of New England is rich in colonial reminiscences; the traveller constantly meeting with objects which remind him of the time when the early colonists were struggling for existence with the Indians; when, relieved from their common enemy, they persecuted one another; when the regicides lay concealed amongst them; when they entered into defensive leagues against their enemies the French, who overhung their northern border; and when they merged into that still mightier league, which embraced the greater part of the Atlantic sea-board, and gave nationality and independence to half a continent.

Between Worcester and Boston the country now looked very different from what it appeared when I first passed over it on my way to Washington. It was then arrayed in the garb of winter, but was now clad in the warmer and more attractive. habiliments of autumn. The trees were beginning to lose their freshness, and some of them had slightly changed their colour; but that transformation had not yet been wrought in them which arrays in such brilliant effects the last stages of vegetation in America for the year. When the frost comes early, the change is sometimes wrought almost in a night. To-day the forest seems clothed in one extended mantle of green—to-morrow, and it appears to have appropriated to itself the celebrated coat of Joseph. The change looks like the work of magic. The leaves are "killed" by the frost during the night, and dyed in their new colours by the sun of the succeeding morning. When a large expanse of it can be commanded by the eye, nothing can exceed in beauty the American forest thus bedecked in its brilliant robe of many colours.

The eastern portion of Massachussetts is very flat, and is in this respect quite a contrast to its western section, lying between the beautiful town of Springfield and the Hudson. The soil is light, and much of it is under pasturage. The vegetation became more stunted as we approached the coast, and we were surrounded by many of the indications which usually mark a tract consisting of a marine deposit.

After remaining a few days in Boston, I proceeded by railway to Lowell, the distance being about twenty-five miles. In point of construction, this line was one of the best on which I had travelled in America. The great majority of my fellow-travellers were New Englanders, and not a few of them would have served as specimens of the genuine Yankee. One cannot fail to observe the tone and demeanour which distinguish the population of this part of the country from that inhabiting the south and west. They are sober, sedate, and persevering; not restless and impatient, like their more mercurial fellow-countrymen.

I was seated beside a resident of Bangor, in the State of Maine. Amongst other subjects of conversation we canvassed the merits of the treaty of Washington, by which the perilous question of the north-eastern boundary was settled. In one thing we were quite agreed, viz. in being both displeased with the treaty; he asserting that Mr. Webster should not have given up an inch of ground in Maine, and I contending that Lord Ashburton went very unnecessarily out of his way to cede Rouse's Point to the Republic. Thus, although we both came to the same conclusion, that the treaty was indefensible, we approached it from very different directions. He was on the whole, however, well pleased that the dispute had been peaceably settled. It was not the territory in dispute, he said, that he cared for, but the principle at issue. The land itself was worth nothing, as he illustrated by assuring me that the few who lived in it had, in winter, to be put into warm water in the morning to " thaw their eyes open! " But he did not like the idea of his country being bullied—which he thought she had been—notwithstanding the pains I had been at to show him that for what she lost in Maine, she had received far more than an equivalent elsewhere ; and that peace, on our part, would have been more cheaply purchased by the simple concession of the line contended for as the boundary of the State.

On approaching Lowell, I looked in vain for the usual indications of a manufacturing town with us, the tall chimneys and the thick volumes of black smoke belched forth by them. Being supplied with an abundant water power, it consumes but little coal in carrying on its manufacturing operations, the bulk of that which it does consume being anthracite and not bituminous coal. On arriving I was at once struck with the cleanly, airy and comfortable aspect of the town ; cheerfulness seeming to reign around, and employment and competence to be the lot of all.

The town of Lowell, a creation as it were of yesterday, is situated on the south bank of the Merrimac, close to the junction of the Concord with that stream. Immediately above it are the falls of the Merrimac, known as the Pawtucket Falls, and which supply the town with the motive power for nearly all its machinery. In 1820 Lowell was scarcely known as a village, its population at that time not exceeding 200 souls. It is now, in little more than a quarter of a century, the second city in Massachussetts in point of size and wealth, and about the twelfth in the United States. Its present population must exceed 30,000.

Until recently American manufactures have had a very up-hill game to play. During the colonial times the jealousy of the mother country threw every obstacle in their way. Still they had in them a germ of vitality which not only outlived every effort made to quench it, but which also enabled them to expand, notwithstanding all the adverse influences against which they had to contend. The imperial legislation of the period would be ludicrous if it were not lamentable, redolent as it was of the spirit of monopoly and self-interest. Its whole object was to make the colonist a consumer, and nothing else, of articles of manufacture, confining his efforts at production to the business of agriculture. If a manu-facturing interest raised its head, no matter how humbly, in any of the colonies, it was not directly legislated down, it is true, but was immediately surrounded by conditions and restrictions which, in too many instances, sufficed to cripple and destroy it. The imperial mind seemed to be peculiarly jealous of the manufacture of hats; an epitome of the legislation in respect to which, if now published, would scarcely be credited, were it not that the whole is to be found in the Statutes at Large. Of course, no hats of colonial manufacture were allowed to cover a British head on what was strictly speaking British ground. But not only were the colonists disabled from exporting their hats to England, they were also forbidden to export them to the adjacent colonies. A hat made in New Jersey was not only forbidden the English market; it was also a *malum prohibitum* in that of New York or Massachussetts. And to enhance as much as possible their value in the colony in which they were manufactured, it was forbidden to convey them from point to point by means of horses. In carrying them to market, therefore, the manufacturer had to take as many upon his head or shoulders as he conveniently could; but to the ordinary modes of conveyance for merchandise he could not resort without the violation of an imperial act. This is a mere specimen of the narrow-minded and sordid spirit in which our colonial legislation was so long conceived. If it discovers any consistent object throughout, it was that it might render itself as odious and vexatious as possible to those who were long in a position which rendered any thing but submission hopeless. The wonder is not that the Americans rebelled in 1776, but that they bore the unnatural treatment to which they were subjected so long. It was not the stamp act or the tea tax that originated the American revolution, but that feeling of alienation from the mother country which had been for the previons century gradually taking possession of the American mind. These acts of parliament were but the pretext, not the cause, of the outbreak. The mine was long laid, they only set fire to the train.

Notwithstanding the many difficulties with which they had to contend, colonial manufactures had taken a firm hold on the continent for some time previous to the epoch of the Revolution. That event, by freeing them from all imperial restrictions, and throwing the American people for some time upon their own resources, afforded them an opportunity by which they failed not to profit. The revolted colonies not only emerged from the war with an independent political existence, but also with a manufacturing interest exhibiting itself in unwonted activity at different points, from the sources of the Connecticut to the mouth of the St. Mary's. This interest steadily progressed, with occasional checks, until the war of 1812, when the Republic was once more, as regarded its consumption of manufactured articles, thrown to a considerable extent upon its own resources. So much so was this the case, that large sections of the country, where the maple was not abundant, had to supply themselves with sugar made from the stalk of the Indian corn. During the war, a large amount of additional capital was invested in the business of manufacturing, to which the three years from 1812 to 1815 gave an immense and an enduring stimulus. Still, even as far down as 1816, the manufacturing system in America had attained, as compared with that of England, but a trifling development; the whole consumption of raw cotton by the American looms for that year being but about half that now consumed by those of Lowell alone, and not more than one-eighth the annual consumption of England at the same period. From that time, by adventitious aids, the system has been forced into rapid growth, until it now owns no rival but that of England herself.

But as far down as 1816, Lowell, now the American Manchester, was undreamt of. A few huts then dotted the banks of the Merrimac, but the Pawtucket Falls had no interest but such as arose from their scenic attractions. Indeed it was not until ten years afterwards that the advantages of its site were fully appreciated; immediately on which the capital of Boston was rapidly invested in it. And what has been the result? The town of Lowell, with all its wealth, industry, achievements, and prospects. In twenty years its population increased one hundredfold; the value of its property during the same period was enhanced one hundred and twenty fold. In 1820 its population, as already observed, was about 200; the value of its property not above 100,000 dollars.

In 1840 its population was 20,000, and its property was assessed at 12,500,000 dollars.

It is supplied with motive power by means of a broad and deep canal, proceeding from the upper level of the Falls along the bank of the river; the majority of the mills and factories being built between this canal and the stream. The canal serves the purpose of a never-failing mill-dam to them all, each drawing from it the supply of water necessary for the working of its machinery. The motive power thus placed at the disposal of capital is equal to the task of turning about 300,000 spindles. In 1844 the number in use did not exceed 170,000; there was therefore power then wasted, sufficient to turn 130,000 more. But as new companies are constantly springing up, a power so available will not long be unemployed.

Almost all the mills in Lowell of any great size, are owned by incorporated companies. A few years ago there were eleven such companies, owning amongst them no less than thirty-two mills, exclusive of print and dye-works, and all supplied with power from the canal. The chief of these is that known as the Merrimac Company, which owns most of the valuable property in the neighbourhood. To it belongs the canal itself, the other companies, as it were, renting the use of it. In addition to several large mills, the Merrimac Company possesses a large machine establishment, in which is manufactured the machinery used in most of the other mills. In addition to the mills owned by the companies, there are some factories of a miscellaneous description, and on a comparatively small scale, owned by private individuals. The great proprietary company, from the very first, took good care that the enterprise of others should not seriously compete with it, by purchasing, when it could be procured at a low rate, all the ground on both sides of the river immediately below the Falls. It is in this way that the other companies are not only dependent upon it for their water power, but are also its lessees or grantees, as regards the very sites on which their mills are erected.

In 1844 there were upwards of 5,000 looms at work in the establishments of the companies, who were then employing nearly 10,000 people, of whom only about one-fourth were males. Scarcely any children were employed under fifteen years of age. The average wages of a male were then from seventy-five to eighty cents a day, or about four dollars eighty cents a week, which make about a pound sterling. Those of a female were from thirty to thirty-five cents a day, or about two dollars a week, being 8s. 4d. sterling. In many cases they were higher. The wages here specified were, in both cases, received exclusive of board.

In 1844 the aggregate produce of the different companies amounted to about 60,000,000 yards of cotton. This constituted their produce simply in the shape of plain goods, their print and dye works during the same year turning out upwards of 15,000,000 yards of printed cloth. The consumption of raw cotton was close upon 20,000,000 lbs.; the aggregate consumption of the Union during the same year was nearly 170,000,000 lbs.; so that Lowell, which as late as 1820 had no existence as a manu-facturing town, was consuming, in little more than twenty years after its foundation, fully one-eighth of all the raw cotton manufactured into fabrics in the

United States.   In 1816, as already intimated, the whole consumption of the American looms did not exceed 11,000,000 lbs.   By this time Lowell alone must be consuming nearly treble that quantity.

The operatives in the different establishments are paid their wages once a month, the companies, however, paying their respective workmen on different days, an arrangement which obviously serves more than one good purpose.   A great portion of the wages thus monthly received is deposited in the Savings' Bank, particularly by the females, who make their work in Lowell a stepping-stone to a better state of existence.   After labouring there for a few years they amass several hundred dollars, marry, and go off with their husbands to the West, buy land, and enjoy more than a competency for the remainder of their days.

In all that conduces to the improvement of the physical and moral condition of the operatives, the companies seem to take a common interest, working together to a common end.   The mills are kept as clean, and as well ventilated, as such establishments can be, and their inmates, with but few exceptions, appear in the best of health; nor is there about them that look of settled melancholy which so often beclouds the faces of our own operatives.   They are comparatively light-hearted, their livelihood being less precarious, and their future prospects far brighter, if they will only improve their opportunities, than those of the English factory-labourer.

Every attention is also paid in Lowell to the education, not only of the young, but also of the adults. By economy of their time and means the women not only manage to be instructed in the elementary branches of education, but also to be taught some of the accomplishments of their sex. It would not be easy to find a more acute and intelligent set of men anywhere than are the artizans and mechanics of Lowell. They have established an institution for their mutual improvement, which is accommodated in a substantial and handsome-looking edifice known as Mechanics' Hall. There are other institutions on a smaller scale, but of a kindred nature, in Lowell. It also possesses eight grammar-schools, at which the pupils who attend receive an excellent education. In addition to this it has no less than thirty free public schools, at which the children of the poorer classes are educated. The number of children attending all the schools is about 6,000, and this out of a population of about 30,000. As elsewhere in the Union, the great business of secular education is harmoniously promoted, without being marred and obstructed by sectarian bigotry and jealousy. Even the Catholics, who are numerous in Lowell, join with the Protestants in the work, all parties wisely and properly agreeing to forget their differences, in furthering that in which they have a common interest,—the education of the young.

Such is Lowell, the growth as it were of a night, the quick result of arbitrary minimums; the fondling of Boston capital, and the pet child of American protection. If it does not owe its existence to high tariffs, its unexampled progress is at least attributable to them. Two years after its incorporation as a city, the almost prohibitive tariff of 1828 was passed, which enabled Lowell at once to realise the most sanguine expectations of its projectors. It was no wonder that under the fostering influence of that tariff, the manufactures of America, both at Lowell and elsewhere, rapidly developed themselves, seeing that its effect was to secure by law to capital invested in a particular employment, a much larger profit than it could count upon with any certainty when otherwise employed. The rise of manufacturing communities in other States as well as in Massachussetts has been the consequence, — the manufacturing capitalist finding himself everywhere rapidly enriched by act of Congress at the expense of the consumer. The plethoric corporations of Lowell owing their success to protection, it is no wonder that they should take the lead in its advocacy. When the Compromise bill expired in 1842, they managed to secure the enactment of a tariff more stringent in its provisions, and consequently more favourable to themselves, than that which had existed for the previous ten years. The injustice to the consumer of the fiscal system established in that year became so manifest in 1846, that it was at length overthrown to make way for the revenue tariff of that year. The manufacturers fought hard in its defence, but in vain. Massachussetts took the lead on their side, Lowell led Massachussets, the Merrimac Company led Lowell, and Mr. Appleton led the Company. But the consumers had got their eyes opened, and saw no reason why they should any longer be taxed in addition to what they were willing to pay for the support of the Government, for the benefit of Massachussetts, Lowell, the Merrimac Company, or Mr. Appleton. The fight, however, was a severe one, and if the free-trade party triumphed on the occasion, it was only by just escaping a defeat.

Although Lowell is, perhaps, the spot in which is concentrated the greatest amount of manufacturing energy, and in which the largest investment of capital has been made for the sole purpose of manufacturing, it forms but a single point in the general survey of the industrial system of America. There is scarcely a State in the Union in which manufactures of some kind or other have not sprung up. The system has as yet obtained but a partial development west of the Alleganies, but most of the sea-board States present to the observer numerous points characterised by great industrial activity. Massachussetts is undoubtedly preeminent in the extent to which she has identified herself with manufactures, in the proper acceptation of the term. In 1846 the capital invested in the business of manufacture in that State must have amounted to from fifty to sixty millions of dollars. In 1837 the amount invested was upwards of fifty-two millions, and the value of the manufactures produced was above eighty-five millions. Between that period and 1842, that is to say, during the last five years of the existence of the Compromise Act, there were no great additional investments made, the operation of that Act not being such, as regarded home fabrics, as to induce capitalists to turn their attention extensively to the business of manufacture. At the same time there was great uncertainty as to the commercial policy which would be pursued on the expiration of the Act, which served as an additional drawback to such an investment of capital. But on the passing of the high tariff act of 1842, when the Union in its economical policy appeared to be reverting to the order of things established in 1828, home manufactures being protected against serious competition,

o 2

and manufacturing capital being virtually guaranteed large returns by Congress itself, great additions were made to that capital; so that the amount now employed in Massachussetts cannot fall much short of sixty millions of dollars. Whether the low tariff bill of 1846 has caused any withdrawal of capital, or checked the increase of capital invested in manufactures, I cannot say. But although Massachussetts may thus claim the lead as the chief manufacturing State, she is behind one of the sisterhood of States, at least, in the amount of capital invested in industrial pursuits, in the broader sense of the term.

When the manufacturing districts of the Union are spoken of, the States of New England are generally alluded to. They are six in number, and are all more or less employed in the business of manufacture. Maine, the most northerly of these, has extensive works for the production of cotton and woollen fabrics, together with several paper mills, and cast-iron works. There is also a great quantity of yarn and coarse cloth produced at the houses of the farmers and others, whilst there are numerous establishments throughout the State engaged, each to a small extent, in miscellaneous manufacture. The capital thus invested in Maine in 1846, must have amounted in all to nine millions of dollars. New Hampshire, which lies to the west of it, is, perhaps, better provided with water-power than any other State in the Union. Of this it has already taken advantage to such an extent as to threaten Massachussetts with a formidable rivalry. In Nashua, Dover, and other places, cotton and woollen factories have rapidly sprung up, and there is scarcely a county in the State but presents its own little focus of manufacturing activity. Some of the more far-seeing statesmen of the revolutionary epoch predicted that New Hampshire would yet owe her prosperity chiefly, if not exclusively, to the system of manufactures which would spring up within her limits. She is in full career to fulfil their predictions, and Massachussetts will have a hard struggle to keep up with her more rugged sister. The water-power of Massachussetts is confined to a few localities, whereas, from its broken and mountainous character, that of New Hampshire is diffused throughout its length and breadth.

The State of Vermont, which lies to the west of New Hampshire, is also abundantly supplied with water-power, the same system of mountains traversing them both. The former, however, is far behind the latter State in industrial enterprise. The amount of capital at present invested in manufactures in Vermont, cannot much exceed five millions of dollars. And yet from her position, Lake Champlain bounding her for its whole length on the west, and opening up a highway for her to the north and the south, one would have expected greater things from her in this respect. Passing over Massachussetts, which we have already considered in this connexion, we come to the little State of Rhode Island, the miscellaneous manufacturing industry of which at present employs a capital of about twelve millions of dollars. Cottons and woollens are its chief products; the number of its woollen manufactories having been in 1840 no less than forty-one, and that of its cotton mills two hundred and nine. From the richness of some of the valleys which intersect it, particularly that of the Connecticut, the State of Connecticut is proportionately more extensively engaged in agriculture than any of the other States of New England.   But she is not deficient in manufacturing enterprise, her capital invested in manufactures being about fifteen millions of dollars.

The capital now employed for manufacturing purposes in the six New England States, is upwards of one hundred millions of dollars.

Leaving New England, the State which, both from its position and the extent to which it has engaged in the business of manufacture, first attracts attention, is New York.   It is abundantly supplied with water-power, which has been turned, more or less, to good account, in various districts of the State.   This water-power is not only derived from the rapid changes of level which take place in the channels of most of its rivers, but is in part produced by the Erie canal, the waste water of which, in addition to irrigating and fertilizing the country in many parts where water is much required, supplies the power by which machinery may be driven.   The chief seats of New York manufacture are Rochester and the neighbourhood of Lockport.   The almost inexhaustible water-power with which the rapids and the Falls of the Genesee supply the former place, has as yet been chiefly applied to the manufacture of flour; but factories of different kinds are rapidly springing up in it, and its annual production is now of a very miscellaneous character.   Small arms and tools of all descriptions are produced here to a great extent, and some of the largest tanneries of the State are on the banks of the Genesee.   Above the falls, the water-power supplied by the rapids has been turned to account on both sides of the river, a succession of huge stone edifices, erected for different manufacturing purposes, confronting each other on either bank. But below the upper fall, the two sides of the river have, as at Lowell, been monopolized by those who turn the available water-power to account on but one bank. The mills and factories erected immediately below the fall, on the Genesee, occupy successive sites on its left bank, each being supplied with the power required to drive its machinery from a common canal, which, like the Pawtucket canal, has its origin at the upper level of the fall, and in its course hems the mills in between it and the river. The water drawn from this canal, after turning the wheels of the different mills, falls in numerous cascades down the bank in reaching the lower level. A great power is thus wasted, the water, in some cases, being capable of being used three different times before it attains the level of the stream below the fall. With the exception of one flour-mill, there is no manufacturing establishment on the opposite bank, which is owned in common by the mill owners on the left side of the river, and which cannot be either sold or leased for manufacturing purposes of any kind, without the consent of all. There is a double purpose to be served by this arrangement—to keep down competition, and to prevent too large a draught upon the water-power afforded by the river, which sometimes, during the protracted heats of summer, becomes so low as for a few weeks scarcely to supply sufficient motive power for the establishments on the left bank. But from the rapids above the upper, to the end of those below the lower falls, the volume of the Genesee is capable of being used by different groups of mills and factories, ten times over, before it reaches the level of Lake Ontario. As yet, it is only the upper fall, with the rapids above it, that has been turned extensively to account.

At Lockport, manufactures have taken a different turn from that which they have as yet mainly taken at Rochester. At the former place cloths of different kinds form the chief product of the mills. The coarse cotton fabric, which is known as Lockport Factory, has attained a very wide celebrity, and is extensively consumed, not only on the American side of the lakes, but also in Canada. It is a heavy bodied fabric, and competes successfully not only with English products of a similar texture, but also with those of New England. New York also manufactures paper to a great extent. The whole amount of manufacturing capital employed by her, must now be above sixty millions of dollars. This, however, is employed in the most miscellaneous production, the amount invested in manufactures, in the ordinary sense of the term, being much less in New York than in Massachussetts.

About fifteen miles from the city of New York, stands the manufacturing town of Paterson in the State of New Jersey. It is beautifully situated upon the banks of the Passaic River, a little below the Falls of Passaic, where the river takes a perpendicular plunge of about seventy-two feet. A canal from the upper level supplies the town with the water-power which it uses, a power which has as yet been but partially turned to account. There are a few woollen factories in Paterson, but its chief product is in the form of cotton fabrics of different textures, the number of cotton-mills being about twenty, having nearly 50,000 spindles at work amongst them. The capital invested in manufactures of all kinds in the town, amounts to about two millions of dollars. The town which ranks next in this State, in point of importance, as regards miscellaneous manufacture, is Newark, about nine miles from New York. At Trenton much paper is made. The total amount of capital now employed in manufactures in the State of New Jersey, is but little under thirteen millions of dollars.

As regards manufacturing in its ordinary acceptation, Pennsylvania falls considerably behind both Massachussetts and New York. But if we take the amount of capital invested in industrial pursuits of all kinds in Pennsylvania, exclusive of commerce, and inclusive not only of her mining operations, but also of the amount of money invested in the construction of public works, with a view chiefly to rendering available her enormous mineral resources, that State, if she does not take the lead of all, will certainly fall behind none other in the confederacy. No less than thirty-four millions of dollars have been invested in canals and railways, chiefly designed to facilitate the transportation of coal from the vast coal fields of the State to tide-water. As far back as 1840, Pennsylvania possessed upwards of 100 cotton factories, working amongst them about 150,000 spindles. But it is evident, when we consider the character of her resources, that the manufactures of this State will not, for some time to come at least, enter very seriously into competition with those of New England. The product which will chiefly spring from the manufacturing energy of Pennsylvania, will be iron, in every shape in which it can be produced. She has got the ore in abundance in her hills and mountains, and the fuel in equal abundance required to convert it to practical purposes. The amount of capital now employed in industrial pursuits in Pennsylvania, exclusive of that invested in works mainly designed for the development of the vast mineral resources of the State, is about forty millions of dollars.

The amount of manufacturing capital employed by the little State of Delaware, is about a million and a half of dollars, of which nearly one-fourth is invested in cotton factories, there being eleven in the State, with nearly 25,000 spindles amongst them.

I do not stop here to inquire whether slavery has had anything to do with the retardation of Maryland in regard to manufactures or not, but certain it is that she has not turned her opportunities to the same account as so many of her northern sisters have done with theirs. She is not only abundantly supplied with water-power both by the Potomac and the Patapsco, but both these streams present her with available water-power, close to tide-water. At Harper's Ferry, the power offered by the rapids of the Potomac, both to Maryland and Virginia—for it runs between the two States—is immense; whilst about fifteen miles from Washington the falls of the river afford them both, in almost inexhaustible supply, this great element of manufacturing industry. But both States seemed content to sleep over their opportunities until the adventurous spirit of northern enterprise led parties from the North to purchase the property in the neighbourhood of the falls. It has since, as already mentioned, been laid out into land and water lots, with the no very happy baptism of South Lowell. The advantages of its site will, therefore, not go much longer unimproved. In addition to this there are available rapids on the Potomac at Georgetown, close to Washington and tide-water, as there are also on the Patapsco, about ten miles above Baltimore. The valley of this latter river is the chief seat of Maryland manufacture. About twenty miles above Baltimore are several cotton, woollen, and flouring establishments; whilst some distance lower down the river are iron-works and rolling-mills on a large scale. At the latter, railway iron is now rolled in great quantities. There are from twenty to twenty-five cotton factories in the State, whilst the capital employed in manufactures of all kinds is below eight millions of dollars.

Virginia is also backward in the business of manufacture, as compared with what she might have done in this respect. She has a bountiful share of the water-power common to all the Atlantic States. It is chiefly at Richmond, her capital, that she has as yet taken advantage of it, the manufactures of which have already been alluded to. Her cotton factories do not exceed in number two dozen; the spindles which they have amongst them not amounting to 50,000. Flour and tobacco figure largely amongst the articles of manufacture produced by this State. The total amount of the manufacturing capital of Virginia does not exceed twelve millions of dollars.

Every effort has lately been made by the North to infuse a manufacturing spirit into the Virginians. Not that it was desirous of rearing up any formidable opposition to itself in the South, but that, by rendering Virginia a manufacturing State, the North would secure her vote on all questions affecting protection to home fabrics, an accession of strength which would render it irresistible in the national councils. But the Virginians are in this respect inert materials to work upon, and the North will find it more to its purpose to transfer a portion of its capital to the banks of the James River and the Potomac, than to confine itself to stimulating the Virginians to manufacturing enterprise. Indeed this is being already done; many Northerners having already entered Virginia, with a view to turning its vast and long-neglected resources to account; and it is not unlikely that, ere long, Richmond will be doubled in size, wealth, and importance, by the influx of northern capital to the banks of the James. The Northerner has an additional inducement to the adventure, in the fact that free has been found more profitable in the business of manufacture than slave labour, even in Virginia herself. The experiment has been tried in the immediate vicinity of Richmond. One of the large factories on the opposite bank of the river is entirely worked by white operatives, and the result has told against the system of employing slave labour in the factory. I was interested, considering the latitude in which I then was, to see, on the dinner-bell ringing, crowds of white men, women, and children emerging from the factory, as if it had been in Paterson or Lowell, instead of in sight of Richmond. It speaks volumes of the want of enterprise which characterises the Virginians, that although one of the finest bituminous coal-beds in the Union, both on account of its supply and its availability, is within a few miles from the capital, it is worked by an English company. The largest iron-work in the town is worked by Welshmen, whilst Scotchmen are, or till very lately have been, the chief merchants of the place.

The manufactures of North Carolina are, and ever have been, on a limited scale—coarse cotton cloth, designed for negro wear, being the chief product of her mills, which are upwards of twenty in number, with nearly 50,000 spindles. The whole capital employed by her in the business of manufacture falls under four millions of dollars.

South Carolina employs about the same amount of capital in a similar way, her chief product in the shape of manufacture being also the coarse Osnaburg cloth, in which the negroes are almost exclusively clad. It is generally made of the roughest part of the cotton crop, such indeed as cannot be exported; and as the quantity of the raw material that enters into it is great, as well as its quality inferior, the New England looms cannot compete in the Southern markets with this domestic fabric. The factories of South Carolina, which are all on a small scale, also produce a considerable quantity of yarn. There are likewise about half-a-dozen iron factories in the State. Those engaged in the production of cotton yarn and coarse cloths are not so profitable as they were some years back, but still return a larger per-centage upon the capital employed than is realized by those who are occupied in the production of the great staples of the State. The factories of South Carolina are chiefly confined to its midland district, which is intersected by the ridge of low sand hills already alluded to, from which a never-failing supply of water is procured.

The State of Georgia comes next in order. It has about twenty cotton factories, producing yarns and negro clothing. The amount of capital employed in these and other factories is about three millions of dollars. The profitable character of the coarse cotton manufactures of the South may be appreciated from the fact, that the Richmond factory, in Georgia, established by a joint-stock company in 1833, averaged, down to 1844, an annual profit of 18 per cent., and for two years afterwards 25 per cent. Again, the Columbus factory, established in 1834, paid nothing for the first four years, the parties managing it being confessedly wanting in skill and experience. Since 1838, however, it has well made up for the want of profits for these years, the average profits since that year having been 20 per cent. The Roswell factory has also paid 20 per cent. since 1839, the date of its establishment.

In Alabama, similar establishments have netted 25 per cent. profit, after allowing for bad debts. The capital employed in manufactures in this State is about three millions of dollars.

The cotton manufactures of the State of Mississippi are almost too insignificant to notice. The State applies about two millions of dollars to the purposes of miscellaneous manufacture.

The manufacturing enterprise of Louisiana is principally applied to the production of sugar, which is its great and most profitable product. It has almost entirely abandoned the growth of cotton for the cultivation of the cane. The capital invested in it in manufactures is about eight millions of dollars.

Florida is yet destined to be the active rival of Louisiana in the production of the cane and the manufacture of sugar. But as yet every interest is, in that State, like the State itself, in its infancy.

The total amount of capital now employed for the purposes of manufacture, including that of articles of every kind, in the different States in the valley of the Mississippi, exclusive of the States of Mississippi and Louisiana, is fully forty-five millions of dollars. Of this aggregate amount, Ohio employs the largest share, the capital invested in manufactures in that State alone being eighteen millions. Kentucky comes next, with a capital of six millions. Indiana follows with five millions, and Tennessee and Illinois come next, each with about four; Michigan and Missouri follow, with about three millions and a half and three millions respectively; and they are followed by Wisconsin with about 700,000 dollars; Arkansas with about 500,000; and Iowa with scarcely 200,000. These last, however, are, like Florida, as yet infant States, their different interests having scarcely had time to take a definite shape since their admission into the Union.

It will thus be seen that, in its diversified phases, the industrial, as contradistinguished from the agricultural interest, has widely, if not universally, established itself in America. The chief seats of manufacture, however, are to be found in New England, and in the States of New York, New Jersey, Pennsylvania, and Ohio. We may also here include Maryland and Virginia. Manufactures have as yet taken but a slender hold of the bulk of the valley of the Mississippi, or of the States south of Chesapeake Bay on the Atlantic, and those on the Mexican Gulf. But it is their ubiquity that gives such homogeneity to the protective principle in America. Were they confined to the Northern section of the Union, and the cultivation of the raw material were alone the occupation of the South, we might expect to find the free-trade and protectionist parties separated from each other by a geographical line. But they are not so confined; and small though the manufacturing interest as yet is, in point of numbers and capital employed, in such States as North and South Carolina, Georgia, Alabama, and Mississippi, and

almost all the States in the valley, it manages, in
conjunction with party predilections, to give the protectionists a good footing even in the States whose
chief business is the production of the great staple of
the South. Thus there is many a Whig in the
South who might not be a protectionist, but for the
presence in his own State of an interest to protect, at
the same time that the pressure of that interest upon
him may be so small, the interest itself being comparatively so, that but for being a Whig he would not
yield to it. It is the two combined that throw him
into the arms of the Northern capitalists, running
counter, as he does, in every vote which he gives in
their favour, to the general·interests of the South.
The Southern Whigs feel this double pressure upon
them, that of party and that of a local manufacturing
interest, very irksome and injurious to them politically;
and there are not a few of them who would joyfully
accept of a final settlement of the tariff question,
even were it of the most ultra free-trade kind, provided it were only final.

The following statements will serve to show the
distribution of manufacturing capital and energy
throughout the United States. I take the figures
from Mr. M‘Gregor's invaluable work, entitled " The
Progress of America," to which I am indebted for
many of the statistical illustrations with which I have
endeavoured to show how rapid has been the industrial development of the Union. In 1840, the total
capital invested in manufactures throughout the
United States was close upon 268,000,000 of dollars.
Of this aggregate amount New York alone employed from 55,000,000 to 56,000,000, Massachussetts
42,000,000 and Pennsylvania 32,000,000, in round
numbers. Next in order came Ohio, with from

16,000,000 to 17,000,000 invested in manufacture ; after which followed Connecticut with 14,000,000, New Jersey and Virginia with from 11,000,000 to 12,000,000 each, New Hampshire with 10,000,000, and Maine with upwards of 7,000,000. It is unnecessary to pursue the comparison further. From this it will be seen, that as regards capital invested in manufacture in its most extensive signification, New York took the lead, being followed by Massachussetts, Pennsylvania, Ohio, Connecticut, New Jersey, and Virginia. And this is, perhaps, the order in which they still range, each State having undoubtedly added largely to its capital during the eight years that have intervened. But if we take the term "manufactures" in its stricter and more limited acceptation, we find that the order in which the States follow each other is greatly changed.

Let us see how the case stands with regard to cotton manufactures. The total capital invested in this branch of industry, in 1840, was from 51,000,000 to 52,000,000, about one-fifth of the total capital invested in manufactures generally. Of this aggregate amount, from 17,000,000 to 18,000,000 belonged to Massachussetts alone, from 7,000,000 to 8,000,000 to Rhode Island, nearly 6,000,000 to New Hampshire, about 5,000,000 only to New York, and from 3,000,000 to 4,000,000 to Pennsylvania. New York, which took the lead in the other case, is here only fourth in the scale, the order in which the States stand, in reference to the amount of capital respectively employed by them in the manufacture of cotton being, Massachussetts, Rhode Island, New Hampshire, New York, Pennsylvania, &c. As regards the capital employed in the manufacture of woollen goods, the order is again changed. Massachussetts, however, still retaining the lead.   The total amount of capital invested, in 1840, in this branch of manufacture was, in round numbers, 16,000,000 of dollars.   Massachussetts owned upwards of one-fourth of the whole, New York a little more than one-fifth, Connecticut about an eighth, and Pennsylvania scarcely one-eleventh.   The value of goods produced during the year was nearly 21,000,000 of dollars, the different States producing pretty much in the proportion of the capital employed by them.   The value of the cotton goods produced during the same year was but a little above 46,000,000.

The total amount of capital invested in manufactures of all kinds in 1848, was very nearly 350,000,000 of dollars, being an increase of nearly 100,000,000 of dollars, or about 40 per cent. in eight years.   The capital invested in cotton manufactures last year amounted to about 64,000,000, being an increase of about 12,000,000, or about 25 per cent. during the same period.   This does not look as if the tariff of 1846 was destructive to the American manufacturing interest.   There was employed during the same year in leather manufactures no less than 33,000,000 of dollars.   The value of cotton goods produced in 1848 was close upon 58,000,000 of dollars, being an increase of about 12,000,000, or about 27 per cent. upon the production of 1840.   The increase in the production of woollen goods has not been so great; still the gross produce is greater than in 1840.   The quantity of leather manufactured last year in the United States is valued at 42,000,000.

It was not until after the revolutionary war that

the cotton manufactures of America made any decided progress. All efforts at their establishment previously to that period had been, more or less, failures. Since 1790, however, they slowly progressed until 1816, after which they became very rapidly developed, having increased about sixteen-fold between that year and 1844. The chief exports of American cotton goods have been to the American markets. The export trade of the fabric has exhibited the most violent fluctuations. The exports, in this particular, to Mexico have fallen off greatly since 1839. During that year the Mexican markets absorbed the white and coloured goods of the Union to the value of 1,335,000 dollars, whereas, in 1843, the importation of American cottons of all kinds into Mexico did not exceed in value 198,000 dollars. The exports to Central America have also fluctuated very much, amounting as they did in 1840 to nearly three times as much as in 1843. The trade with Chili, during the same period, also exhibited a decrease. Until 1841, that with Brazil steadily increased, but declined from that year to 1843. The same fluctuation is discernible if we take the aggregate exports of cotton goods from 1826 to 1843. They were higher in 1831 than in 1843. It was in 1841 that they reached the highest point, being then in value from 12,000,000 to 13,000,000 of dollars. In 1842 they declined to a little over 9,000,000, and in 1843, to below 7,000,000.

The woollen manufactures of America have progressed but slowly as compared with those of cotton. They are almost exclusively produced for home consumption, the quantity of woollen goods exported being exceedingly small.

The manufacture of silk has also made considerable

progress in the United States. Some years ago there was a great deal of speculation in connexion with this branch of industry. Millions of dollars were invested in mulberry-trees, with a view to the culture of silk, the belief having taken possession of the public mind that the silkworm could be reared in America from Maine to Georgia. The mania did not last long, but much money was lost during its prevalence. Since that time the silk manufacture of America has remained almost stationary, having enjoyed for some years afterwards rather a bad reputation. The State of Ohio produces a good deal of silk, specimens of which I have frequently seen. It is, as well may be supposed, a very inferior article, and it will be long ere America produces any silk fabrics to which a more flattering epithet can be applied.

The iron manufactures of America have already been cursorily alluded to, in treating of the mining interests of Pennsylvania. The total capital invested in connexion with the working of iron, including mining, casting, forging, &c., is upwards of 25,000,000 of dollars. In lead, more than 2,500,000 are invested, which capital is chiefly employed in working the mines at Galena.

Paper forms a not unimportant item in the sum total of American manufactures. The capital employed in producing it is upwards of 5,000,000.

In the manufacture of flour and oil, and in the sawing of timber, upwards of 75,000,000 of dollars are invested. The number of barrels produced per annum by the different flouring-mills in the country, is from 8,000,000 to 9,000,000. The mills themselves are nearly 5,000 in number; whilst of saw-mills there are upwards of 30,000 in the United States.

In estimating the manufacturing facilities possessed

by the United States, many put foremost in the catalogue its almost infinite water-power. But there are others who believe that, for factories producing most classes of goods, steam, where it can be generated at little cost, is preferable to water-power. This may be all very true as regards very large mills, requiring heat for certain processes, which heat may be obtained from the steam after it has served its purpose in driving the machinery ; but it is evident that but for the water-power in which the country abounds, the great bulk of the small factories occupying remote positions would not have had an existence. Steam, even in the most favourable localities for generating it, may be more expensive than water as a simple motive power ; but there is this in favour of steam, that the factory employing it can be built where everything required about an establishment of the kind may be had readily and cheaply. There are many factories now employing steam, in the immediate vicinity of good available water-power. One of the largest manufacturing establishments in America, that known as the Gloucester Mills, situated on the New Jersey bank of the Delaware, a little below Philadelphia, employs steam as its motive power. It is the consideration, that even in America, where water-power is so abundant, steam may be advantageously employed in the business of manufacture, that leads one to anticipate for Philadelphia, which is so favourably situated for a supply of coal, the destiny of being yet the greatest manufacturing emporium of the continent.

But enough has been said to show how extensive and varied is the manufacturing interest of the Union. It is an interest which has in itself all the essential elements of progression; and which will yet, in its onward course, attain a momentum which will enable it to dispense with the adventitious props for which it is now so clamorous in the way of protection. The germ has been, as yet, but laid of the manufacturing system which is destined to permeate America; and if we are to judge of its future progress from its past achievements, the time cannot be far distant ere it attains a colossal magnitude.

It has, therefore, not been a weak interest against which the agriculturists, including the cotton-growers, have had to struggle. Not that the manufacturers are as strong in point either of numbers or of capital as the agriculturists, but they are combined and work together; whereas the agriculturists generally exhibit a want of combination and of a common understanding with one another, when it is most important for them to have both. Not a little of the political success of the manufacturers is attributable to their superior shrewdness, adroitness, and perseverance. If the two classes are measured by the extent of their interests, the agriculturists will be found to eclipse their rivals. In the six States of New England, together with New York, Pennsylvania, New Jersey, Delaware, and Maryland, this is certainly not the case, the value of the annual manufactures of these States considerably exceeding the value of their annual agricultural produce. But if we take the remaining nineteen States of the Union, the value of their agricultural produce so far exceeds that of their manufactures, that, taking all the States together, the balance in point of interest is largely with the agriculturists. Thus the crops produced last year by the States first named, amounted in value to about

216,000,000 of dollars. Their gross manufactures are valued at 252,000,000. This leaves a balance of upwards of 30,000,000 in favour of the manufacturer. But in the other States the crops produced are valued at 356,000,000, whereas the value of their manufactures does not exceed 90,000,000. This leaves a balance of upwards of 260,000,000 in favour of the agriculturist. The value of the whole crop of the Union is thus above 560,000,000, that of its gross manufactured products a little above 340,000,000, leaving, on the whole, a balance in favour of the agriculturist of no less than 220,000,000. In addition to this, the value of agricultural exports far exceeds that of manufactured articles exported. In 1840 the value of all the exports of the Union did not exceed 113,000,000 of dollars. Of this sum no less than 92,000,000 represented the value of agricultural products exported. Last year, the value of the aggregate exports reached the enormous amount of 154,000,000. Of this a still larger proportion was the value, exclusively, of agricultural productions. Whatever may be the fate of the export trade as regards manufactures, that in connexion with produce is destined largely and rapidly to increase. It is therefore the great source of wealth to the country. It seems singular, therefore, that the agricultural interest should have suffered itself to be so frequently sacrificed to its less important rival. But the dazzling vision of an " American system," with national self-dependence, sufficed for a long time to mislead the agricultural mind as to its true interests.

I have observed, in a previous chapter, that the interests of the commercial classes are as much identified with free trade as are those of the agriculturists. To this the commercial classes in Boston certainly offer an exception, and this exception has been frequently forced upon the farmers as a proof that all classes in the community had a common interest in a high tariff—in other words, in protection. It is quite true that the leading merchants in Boston generally side with the manufacturers; but it would be erroneous thence to infer that the commercial classes of the Union are identified with them either in feeling or in interest. The leading Boston merchants are peculiarly situated, either having themselves shares in the manufacturing establishments at Lowell and elsewhere, or having fathers, mothers, brothers, sisters, aunts or uncles that have. They are thus more or less in the same boat with the manufacturers; and the same may indeed be said of the agricultural classes of the State of Massachussetts, who are extensively employed during the winter, as already intimated, in the rather incongruous occupation of making boots and shoes. This enables Massachussetts to exhibit a wonderful unanimity on the subject of protection, farmers, manufacturers, merchants and ship-owners all appearing to clamour for it. This seeming identity of interest has imposed not a little upon the farmers elsewhere, who did not take into account the peculiar position of parties in Massachussetts.

A remarkable exception in this respect to the majority of the leading merchants of Boston was presented in the person of Mr. Philip Homer, himself for a long time extensively engaged in that city as an importer on a large scale. He was, fortunately, untrammelled by any connexion with the manufacturers, which, combined with his quick perception and strong good sense, enabled him to take a clear and unbiassed view of the general interests. He was an ardent free - trader, and on retiring from business, went through the country, doing everything in his power to disseminate his views. In 1846, when the tariff bill was under discussion, he procured the use of one of the committee-rooms of the capitol, in which to exhibit rival patterns of British and American manufacture. He put patterns of equal texture together, showing the difference between their prices; and patterns of equal price together, showing that between their textures. Partly to neutralize the effect produced by this exhibition, which was anything but favourable to the pretensions of the American manufacturer, and partly to overawe Congress by a great practical argument, which they hoped would have more weight than those of a mere speculative kind, they determined on holding their ordinary annual fair that year at Washington. Preparations were accordingly made on a most extensive scale for the exhibition. A temporary wooden building was erected for the purpose, in the form of a T, its area being about double that of Guildhall. To this, goods of all kinds, exclusively the produce of domestic industry, were forwarded. In the course of about ten days it was filled with articles of different descriptions, and thrown open to the public. The display was imposing in the extreme, and he would be as bold as he would be unfair who would deny that it was most creditable to American enterprise and skill. But it failed in producing the desired effect. As regarded some, it had the contrary effect to that intended to be produced, for they thought that an

VOL. III.   P

industry which produced such excellent fruit required
no protection to enable it to maintain its ground.
Congress was neither overawed nor convinced—the
tariff bill passed, and shortly afterwards the manufactures of the country had to submit to competition.
Bands of music attended at the exhibition, and everything was done to render it as attractive as possible.
At one time it was intended to bring some of the factory girls from Lowell to it, as specimens of native
produce, but the intention was speedily abandoned.
Multitudes flocked from all parts of the country to
witness the Fair, and Washington was literally glutted
with strangers. This admirably served the purpose
of Mr. Homer, who was all the time proceeding with
his quiet, unobtrusive, but rival exhibition in the
capitol. I was with him one day, when a very fiery
and uncombed young man entered, and after fuming
about for some time, began to attack Mr. Homer in
a no very courteous manner for his enmity to domestic
industry. It was soon evident that he regarded him
as a European—in fact, as an interested agent of the
English manufacturer. Mr. Homer put him right
on this point, informing him that he was a Bostonian
and his fellow-countryman ; but this, instead of paci-
fying, made him all the more furious. It was quite
bad enough for the foreigner, be intimated, thus to
beard the home manufacturer in the very heart of the
Union ; but for a native to do it was something, in
his opinion, worse, if possible, than sacrilege.

"I'm Southerner," he said at last, bursting into
a fit of uncontrollable passion, "and I'll make the
Union yet ring with your name." Having said this,
he left the room, and repaired to the fountain hard
by to cool himself. I asked, when he had left, what

he meant by saying he was " Southerner;" when I was informed that he contributed letters to some Southern newspaper, under that signature.

When the American manufacturers talk of self-dependence as the proper attitude of the Republic, do they mean that it should cut itself off from all commercial intercourse with the world? Yet this is what they must do to realize their dreams. The primary condition to mutual trade is mutual dependence. If America can be brought to a point at which she will want nothing from the rest of the world, the condition will be wanting to her trading with the rest of the world. Trade cannot be all on one side. She may have much to give away, but unless she takes something, the produce of others' industry, in return for it, she cannot dispose of it, unless she do so gratuitously. And should this self-dependence ever be attained, and this national isolation secured, what will become of the shipping interest and the national marine of America? Let the Americans remember that they are much more dependent for the manning of their navy upon a flourishing commercial marine than we are. Indeed, wages for civil employments are so high in America, that this is the only source to which they have to look for the material with which to man their navy. In Europe it is otherwise. The condition to a flourishing commercial marine is a flourishing foreign trade. The pivot on which the foreign trade of America now turns is its export of cotton. Let the manufacturers have their way, and this trade is ruined.

If the manufacturers would only wait patiently for the *denouement*, that which they are so anxious to precipitate will, in all probability, ere long unfold

itself as the natural result of the progress of manufactures in America.   Their water-power is inexhaustible, their machinery will be gradually perfected,
their skill will increase, and the cotton will continue
to be cultivated almost at their very doors.   The
only condition to a complete monopoly of the American market, in which they will long be wanting, is
cheap labour.   But there are facilities in their way
which, if properly turned to account, may more than
compensate them for continued high wages.   By
attempting to realize at once the monopoly which
appears yet to be in store for them, they bring themselves into angry collisions with other interests, upon
the development of which they trench, by seeking to
force the growth of their own, no matter at what
cost to the country.   They can only now monopolize
the home market at a heavy cost to every other
branch of domestic industry—in other words, they
can only protect themselves by the imposition of a
heavy tax for their exclusive benefit upon the great
body of consumers.   Protection thus cuts both ways.
It injures the foreigner, and also the domestic consumer.   Between the two parties thus treated, stands
the protected interest, which alone receives the benefit
of the false policy on which it flourishes.

CHAPTER XI.

AMERICAN CHARACTER.—PHYSICAL CONDITION OF SOCIETY IN AMERICA.

The American Character the reverse of gloomy and morose.—Sensitiveness of the American people. — Its explanation and its excuse.—The Americans more sensitive at home than abroad.—Why they are so, explained.—They are more boastful abroad than they are at home.—This also explained.—Allowances to be made. —The American has cause to feel proud of his Country's Progress.—The national feeling in America resolves itself chiefly into a love of Institutions.—Identification of the American with his political system.—Impossibility of the establishment of Monarchy in the United States.—Stability of Democracy in America.—Monarchy impossible, even in Canada, in the event of its separation from the mother country. --The American's faith in his Country's Destiny.—Influence of this on his feelings and character. —Feeling cherished towards England.—Love of titles in America.—Love of Money.—Fondness for Dress.—Physical condition of society in America.

MANY Europeans quit the shores of the Republic with unfavourable impressions of American character, in the broadest acceptation of the term. But in the majority of instances, those who do so enter the country with preconceived notions of it, and leave it ere they have learnt to discern objects through the right medium. The Americans as a people, for instance, are characterised by some as gloomy and reserved; whereas, if properly approached, they are frank, communicative, and not unfrequently even mercurial in their dispositions. Any one who has mingled much in American society must have seen that gloom was far from being its predominant characteristic, at least in the case of American women. If they have

any fault in this respect as a class, it is not that
of coldness and reserve, but of over vivaciousness,
and a tendency to the frivolous and amusing.   In
parts of the country, where fanaticism in religion
has for some time prevailed, a settled gloom may
be discerned on the majority of countenances;
but it does not so much indicate a morose spirit,
as a real or affected habit of looking serious.   From
a pretty long and intimate acquaintance with American society in most of its phases, I can confi-
dently say, that the traveller who finds the people of
America habitually keeping him at a distance, and
otherwise treating him coldly, must be himself chiefly
to blame for the reception which he experiences.
During my peregrinations through the Union—and
they were many and long—I had frequent opportunity of seeing how English travellers demeaned
themselves on passing through the country.   I invariably found that those who met the Americans
frankly and ingenuously, were treated with the utmost
kindness and warmheartedness, and were consequently
favourably impressed with the character of the
people; whereas, such as travelled through the
country as if it were a compliment to the Republic
that they touched its democratic soil, and as if the
mere fact of their being Englishmen entitled them to
treat all who came in their way with ill-dissembled
hauteur and contumely, were left to find their way
as they best could, the cold shoulder being turned
to them wherever they went.   This is not done from
any feeling of vindictiveness towards them, for they
are generally laughed at on assuming insolent airs
and demanding extra attentions.   Those who will not
treat them frankly, the Americans will not put them-

selves out of their way to receive kindly, nor will they give their confidence to such as expect to gain it without an equivalent. But be frank, fair, and honest with them, treating them not with marked deference, but with ordinary courtesy, and a more kind-hearted, accessible, hospitable and manageable people are not to be found.

The Americans are almost universally known to be a sensitive people. They are more than this; they are over-sensitive. This is a weakness which some travellers delight to play upon. But if they understood its source aright, they would deal more tenderly with it. As a nation, they feel themselves to be in the position of an individual whose permanent place in society has not yet been ascertained. They have struggled in little more than half a century into the first rank amongst the powers of the earth; but, like all new members of a confined and very particular circle, they are not yet quite sure of the firmness of their footing. When they look to the future, they have no reason to doubt the prominency of the position, social, political, and economical, which they will assume. But they are in haste to be all that they are yet destined to be; and although they do not exact from the stranger a positive recognition of all their pretensions, they are sensitive to a degree to any word or action on his part which purports a denial of them. It must be confessed that this weakness has of late very much increased. A sore that is being constantly irritated will soon exhibit all the symptoms of violent inflammation. The feelings of the American people have been wantonly and unnecessarily wounded by successive travellers who have undertaken to depict them, nationally and indi-

vidually, and who, to pander to a prevailing taste in
this country, have generally viewed them on the
ludicrous side. It is a mistake to fancy that the
Americans are impatient of criticism. They will
submit to any amount of it that is fair, when they
discover that it is tendered in an honest spirit. What
they most wince at is the application to them and
their affairs of epithets tending to turn them into
ridicule. You may be as severe as you please with
them, even in their own country as well as out of it,
without irritating them, provided it appears that
your intention is not simply to raise a laugh at their
expense. When I first went to Washington I was
cautioned by one who knew the Americans well, not
to suppress my real sentiments concerning them, but
to be guarded as to the terms and the manner in
which I gave utterance to them. They have been so
frequently unjustly dealt with by English writers,
that they now suspect every Englishman of a predetermination to treat them in a similar manner. I acted
upon the advice which I received, and for the six
months during which I resided in the capital,
I freely indulged in criticism of men and things,
without, so far as I could ascertain, giving the
slightest offence to any one. But there are cases in
which a look, a shrug of the shoulder, or a verbal
expression, may cause the greatest irritation. In this
country it is difficult to understand this sensitiveness
on the part of the American people. England has
her fixed position in the great family of nations, and
at the head of civilization——a position which she has
long occupied, and from which it will be some time
ere she is driven. We care not, therefore, what the
foreigner says or thinks of us. He may look or

express contempt as he walks our streets, or frequents our public places. His praise cannot exalt, nor can his contempt debase us, as a people. The desire of America is to be at least abreast of England in the career of nations; and every expression which falls from the Englishman showing that in his opinion she is yet far behind his own country, grates harshly upon what is after all but a pardonable vanity, springing from a laudable ambition.

The Americans are much more sensitive at home than they are abroad. Their country is but yet young; and when they hear parties abroad who have never seen it, expressing opinions in any degree derogatory to it, they console themselves with the reflection that the disparaging remark has its origin in an ignorance of the country, which is judged of, not from what it really is, but simply as a State of but seventy years' growth. Now in Europe it is but seldom that seventy years of national existence accomplishes much for a people. It is true that more has been done for mankind during the last seventy than perhaps during the previous 700; but the development of a nation in Europe is a slow process at the best, as compared with the course of things in this respect in America. The American, therefore, feels that, if the European would suspend his judgment until he saw and heard for himself, it would be very different from what it is when begotten in prejudice and pronounced in ignorance. This takes the sting from such disparaging criticism abroad as he may chance to hear. But if it is offered at home, unless it is accompanied with all the candour and honesty in which such criticism should alone be indulged in, he has no such reflection to take refuge in, and it wounds him to the quick. If, notwithstanding all the evidences which the country affords of unexampled prosperity, universal contentment, social improvement and material progress, the foreigner still speaks of it, not in terms of severity, but in those of contempt—in terms, in short, which the American feels and knows are not justifiable—he can only refer the criticism to a predetermination to turn everything into ridicule, and is consequently not unjustly offended. Such, unfortunately, is the predetermination with which a large proportion of English travellers in America enter the country, demeaning themselves, during their peregrinations through it, with an ill-disguised air of self-importance, unpalatable to a people who have become jealous from unmerited bad treatment. The consequence is, that every Englishman in America is now on his good behaviour. He is not regarded as candid until he proves himself the reverse, but as prejudiced and unfriendly until he gives testimony of his fairness and honesty.

If the Americans are more sensitive at home than they are abroad, they are more boastful abroad than they are at home. The one is a mere weakness, the other frequently an offence. Many in Europe judge of the American people from the specimens of them who travel. There are, of course, many Americans that travel, who, if they partake largely of the national vanity attributed to them all, have the tact and the courtesy to conceal it. Indeed, some of the best specimens of Americans are, for obvious reasons, those who have travelled much from home. But the great mass of American travellers enter foreign countries with as thick a coat of prejudice about them, as Englishmen generally wear in visiting America. The consequence is that they commit the fault abroad, at which they are so irritated when committed in regard to themselves by the foreigner in America. With the American abroad, however, this fault assumes the reverse phase of that taken by it when committed by the foreigner in America. The Englishman, for instance, who is disposed to view everything in America through a jaundiced eye, and to draw invidious comparisons between the two countries, exalts his own by running down the other. The American, on the other hand, having the same object in view, approaches it from the opposite side, drawing comparisons favourable to his country, not by disparaging others, but by boasting of his own. This may be the weaker, but it is certainly the less offensive manifestation of a common fault. It would be erroneous to suppose that the national vanity which so many Americans exhibit abroad, is prominently manifested at home. At all events it is not obtruded upon the stranger. The evidences of the country's greatness, both present and prospective, are before him when in the country; and to recapitulate them to him under these circumstances would be but to tell a tale twice over. If he does not draw favourable conclusions from what he sees, it is hopeless to expect him to do so from anything that he could hear. The American may be amazed at his real, or annoyed at his wilful blindness, but he generally leaves him to his own inferences. It is only abroad, and when in contact with those who have not had ocular demonstration of it, that he is prone to dwell in a vaunting spirit upon his country's greatness.

Some allowance, however, should be made for the American, even in his most boastful humour. If he

has nothing in a national point of view to be vain of,
he has certainly much of which he ˙can and should
feel proud.   There is no other country on earth
which in so short a time has accomplished so much.
It has but just passed the usual term allotted as the
period of life to man, and yet it takes rank as a first-
rate power.   But let it not be supposed that all this
has been achieved in seventy years.   The American
republic has never had a national infancy, like that
through which most European nations have passed.
The colonies were, in a measure, old whilst they were
yet new.   They were as old as England herself
in point of moral, and new only in point of material,
civilization.   They were not savages who laid the
foundations of our colonial dominion in America, but
emigrants from a highly civilized society, carrying
with them all the moral results of centuries of social
culture. The youth of Anglo-Saxon America was not a
period of barbarism ; its civilization, morally speaking,
was up with our own when it was first colonized.   If
it did not always keep up with it, the reason is to be
found in the nature of the circumstances in which it
was placed.   The civilization of England in the
seventeenth century was transplanted to a country
resembling England in the first.   The barbarism of
nature was a drawback to the rapid development of
the civilization which had been transferred to it.   A
war between the two immediately ensued, the result
of which was the subjugation of the wilderness, and
the civilization of external nature.   But during the
progress of the conflict, particularly in its earliest
and severest stages, the career of intellectual and moral
improvement was necessarily retarded.   The merit
of the American colonists consisted in this, that their

retardation was not much greater and more prolonged. The same conflict is now being waged in the Far West, society there at the present day being the counterpart of what society was on the sea-board colonies two centuries ago. In the colony material civilization had greatly progressed previously to 1776. When, therefore, the independence of America was proclaimed, the country had made large advances in the career of social and material improvement, so that when it became invested with a distinct and separate nationality, it was already comparatively old. The present development of America cannot, then, be regarded as the result of its efforts during the brief period of its independence. The sources of that development are traceable not only back to colonial times, but also to the successive stages of English civilization, long before the colonies were dreamt of. Although the American cannot, thus, refer all his country's greatness to the period of its independence, there is no question that the strides which it has taken during that period cast all its previous advances into the shade. In these he has undoubtedly cause for national pride and self-gratulation.

Intimately connected with the pride of country which generally distinguishes the Americans, is the feeling which they cherish towards their institutions. Indeed, when the national feeling of an American is alluded to, something very different is implied from that which is generally understood by the term. In Europe, and particularly in mountainous countries, the aspect of which is such as to impress itself vividly upon the imagination, the love of country resolves itself into a reverence for locality irrespective of all other considerations. Thus the love which a Swiss bears to his country is attached to the soil constituting Switzerland, without reference to the social or political institutions which may develop themselves in the cantons. And so with the Scottish mountaineer, whose national attachments centre upon the rugged features of his native land. It is seldom that the national feeling exhibits itself to the same extent in the breast of one born and bred in a country surpassingly rich, perhaps, in all the productions which minister to the comforts of life, but destitute of those rough and stern features which so endear his country to the hardy mountaineer. It is quite true that inspiriting historic associations may frequently produce feelings of national attachment similar to those inspired by a grand and imposing development of external nature : it is thus that some of the most patriotic tribes on earth are the inhabitants, not of the rugged mountain defile, but of the rich and monotonous plain. But the American exhibits little or none of the local attachments which distinguish the European. His feelings are more centred upon his institutions than his mere country. He looks upon himself more in the light of a republican than in that of a native of a particular territory. His affections have more to do with the social and political system with which he is connected, than with the soil which he inhabits. The national feelings which he and a European cherishes being thus different in their origin and their object, are also different in their results. The man whose attachments converge upon a particular spot of earth, is miserable if removed from it, no matter how greatly his circumstances otherwise may have been improved by his removal; but give the American his institutions, and he cares but little where you place him. In some parts of the Union the local feeling may be comparatively strong, such as in New England; but it is astonishing how readily even there an American makes up his mind to try his fortunes elsewhere, particularly if he contemplates removal merely to another part of the Union, no matter how remote, or how different in climate and other circumstances from what he has been accustomed to, provided the flag of his country waves over it, and republican institutions accompany him in his wanderings.

Strange as it may seem, this peculiarity, which makes an American think less of his country than of the institutions which characterise it, contributes greatly to the pride which he takes in his country. He is proud of it, not so much for itself as because it is the scene in which an experiment is being tried which engages the anxious attention of the world. The American feels himself much more interested in the success of his scheme of government, if not more identified with it, than the European does in regard to his. The Englishman, for instance, does not feel himself particularly committed to the success of monarchy as a political scheme. He will support it so long as he is convinced that it conduces to the general welfare; and, judging it by this standard, it is likely that he will yet support it for a long time to come. He feels his honour to be involved in the independence of his country, but does not consider himself to be under any obligations to prove this or that political system an efficient one. The political scheme under which he lives he took as part and parcel of his inheritance in a national point of view, and his object is to make the best of it. It is very different, however, with the American. He feels himself to be implicated, not only in the honour and independence of his country, but also in the success of democracy. He has asserted a great principle, and feels that, in attempting to prove it to be practicable, he has assumed an arduous responsibility. He feels himself, therefore, to be directly interested in the success of the political system under which he lives, and all the more so because he is conscious that in looking to its working mankind are divided into two great classes—those who are interested in its failure, and those who yearn for its success. Every American is thus, in his own estimation, the apostle of a particular political creed, in the final triumph and extension of which he finds both himself and his country deeply involved. This gives him a peculiar interest in the political scheme which he represents ; and invests his country with an additional degree of importance in his sight, as in that of many others, from being the scene of an experiment in the success of which not only Americans but mankind are interested. Much, therefore, of the self-importance which the American assumes, particularly abroad, is less traceable to his mere citizenship than to his conscious indentification with the success of democracy. Its manifestation may not always be agreeable to others, but the source of his pride is a legitimate and a noble one. It involves not only his own position, but also the hopes and expectations of humanity.

It is this feeling which renders the establishment of monarchy an impossibility in the United States. The American not only believes that his material interests are best subserved by a democratic form of government, but his pride is also mixed up with its maintenance and its permanency. It is a common thing for Europeans to speculate upon the disintegration of the Union, and the consequent establishment, in some part or parts of it, of the monarchical principle. These speculations are generally based upon precedents, but upon precedents which have in reality no application to America. The republics of old are pointed to as affording illustrations of the tendencies of republicanism. But the republics of old afford no criterion by which to judge of republicanism in America. The experiment which is being tried there is one *sui generis*. Not only are the political principles established different from those which have heretofore been practically recognised ; but the people are also in a better state of preparation for the successful development of the experiment. The social condition of the ancient republics was as different from that of America as night is from day. The political superstructures which arose in them conformed themselves more or less to the nature of their bases. The result was not republicanism, but oligarchy. All that can be said of these so-called republics is, that they were not monarchies. But it does not follow that they were republican. The elementary principle of republicanism is, that government, to be stable, must be deeply rooted in the public will. The governments of the older republics were not so, and they perished—as all usurpations will and must do. The more modern republics, again, are divisible into two classes—such as were assimilated in the principles and in the form of their government to the more ancient, and such as too hastily and inconsiderately assumed the true democratic type. If the former shared the fate of the older republics, it was because

they resembled them in the faultiness of their construction. If the latter were evanescent, and speedily
relapsed into monarchy, it was but the natural result
of hasty and violent transition.   But the mistake lies
in arguing from these cases, particularly the latter,
in our speculations as to the future of America.   It
is but natural that a people who have been for ages
inured to monarchy, whose sentiments are more or
less intertwined and whose sympathies are bound up
with it, should, after having been for a season, either
through their own madness or through the folly of
others, divorced from it, revert to it again on the
first favourable opportunity.   But in doing so they
are only following the true bent of their inclinations,
to which their inconsiderate republican experiment
in reality did violence.   Generations must elapse ere
a people trained and educated to monarchy can be
really converted into republicans : in other words,
a people cannot be suddenly or violently diverted
from that to which they have been trained and accustomed. This is a very simple rule : but simple though
it be, it is precisely that which Europeans overlook
in judging of the stability of democracy in America.
The American Republic, in the first place, differs
essentially from all that have preceded it in the principles on which it is founded : it is not a republic in
simply not being a monarchy : it is a Democratic
Republic, in the broadest sense of the term.   If it is
not a monarchy, neither is it an oligarchy.   It is the
people in reality that rule ; it is not a mere fraction
of them that usurps authority. The success of the
American experiment depended, as it still depends,
upon the character of the people.   As already shown,
the stability of the republic is intimately identified

with the enlightenment of the public mind—in other words, with the great cause of popular education ; it is to the promotion of education that it will in future chiefly owe its success. But its maintenance at first was mainly owing to the political antecedents of the people. It is quite true that they were converted in a day from being the subjects of a monarchy into the citizens of a republic. But let us not overlook the long probation which they underwent for the change. From the very foundation of the colonies, the subjects of the British crown in America were being practised in the art of self-government. The charters which most of the colonies received from the crown were of the most liberal description, and, in fact, constituted the seeds of the future Republic. Prerogative ran high at home in the days of the Restoration ; but so liberal was the charter which Charles II. conceded to the colony of Rhode Island, that from 1776 down to 1842 it served the purposes of a constitution in the State of Rhode Island. The political transition, therefore, which took place in 1776, so far from being a violent one, was but the natural consequence of the political education to which the American colonists had been subjected for a century and a half before. The moment they separated themselves from the imperial crown, they assumed the republican form of government, not from impulse or enthusiasm, but from the very necessity of the case. They had been long taught the lesson of self-reliance and self-control ; and if, so long as they were colonists, they remained monarchists, it was more from old associations and ties than from not being ripe for a republic. The establishment of the Republic in America in 1776, then, was not a violent act, but a necessary one, after the disruption between the colonies and the mother country. This is what those forget who predict that the Republic will speedily relapse into monarchy. But it is in this that consists the essential difference between the American Republic, and the European republics of a modern date. Had the establishment of the Republic in 1776 warred with the habits or done violence to the feelings of the people, its over-throw might have been speedily looked for. But so far was this from being the case, that no other form of government could have been instituted that would have outlived a lustrum. The establishment of a permanent monarchy was as impossible in America in 1776, as was the establishment of a permanent republic in France in 1848. In the one case, the tendency of the people to revert to that to which they were educated, trained, and acccustomed, would have overpowered the system temporarily established amongst them—as it is speedily destined to do in the other. The safety of the American Republic con-sists in this, that in establishing it the American people were not suddenly or violently diverted from the political order of things to which they had been accustomed. Let parties well consider this before they indulge in sinister predictions as to the insta-bility of the political institutions of America. If the Americans have been successful as republicans, it is because they underwent a long probation to the prin-ciple of Republicanism. Under the shadow of a powerful monarchy, to which they belonged, but by which they were really not governed, they practically acquainted themselves with the art of self-govern-ment. The colonies were thus practically republics before they became independent. Institutions, to be stable, must conform to the tastes, habits and genius of a people. Monarchy could not have done so in America in 1776. Republicanism alone was suited to the character of the people and the exigencies of the country. Republicanism alone, therefore, was possible.

It is equally so at the present day. Consider the Americans now—and what is there in their character, feelings, or circumstances to lead them back to monarchy? Everything connected with them tends the other way. Their associations are all republican—their principles and practice have ever been so—their interests have been subserved by republican institutions, and their pride is now involved in their maintenance and extension. The circumstances of the country, and the character and genius of the people, are as much now as in 1776 inimical to monarchy. On what, therefore, rests the supposition so often hazarded by parties in this country, that violence will be done, and that ere long, to the Republic in America? Unless the people can be persuaded to do violence to their feelings, tastes, habits, and associations, and to adopt institutions incompatible with their position and circumstances, there is no fear of democracy in America.

Many point to the accumulation of wealth as that which will work the change. It is quite true that some of the millionnaires of America would have no objection to the establishment of a different order of things. But both in numbers and influence they are insignificant, as compared with the great mass even of the commercial and manufacturing communities, who are staunch democrats at heart. Much more are they so when we take the great agricultural body of America into account. Here, after all, is the stronghold of democracy on the continent. However it may be undermined in the town, its foundations are deeply and securely laid in the township. No one who has mingled much with the American farmers can entertain any serious doubts of the stability of democracy in America. Even were the entire commercial and manufacturing community otherwise disposed, they could make no impression against the strong, sturdy, democratic phalanx engaged in the cultivation of the soil. But the great bulk of the commercial and manufacturing classes are, as already intimated, as devoted to the republican system as any of the farmers can be. During the whole of my intercourse with the Americans, I never met with more than two persons who expressed a desire for a change. One was an old lady who had got a fright at an election, and the other was a young lieutenant in the army, who lisped, through his moustache, his preference for a military despotism to a republican government. It was very evident that he understood neither the one nor the other.

The following will serve to illustrate how deeply the republican sentiment has infused itself into the minds of all classes in America. On my return to Liverpool I visited Eaton Hall, near Chester, in comnany with some Americans who had been my fellow-voyagers. After inspecting the interior, we strolled along the magnificent grounds which enclose that noble pile. One of the company was a retired merphant of New York, who had amassed a large fortune, and occupied a fine mansion in the upper and fashionable part of Broadway. After waiting until he had seen all, I asked him what he thought of it. He replied, that he would give me his opinion when we were in the streets of Chester. I understood his meaning, and asked him if he did not think that the same diversities of light and shade would soon be exhibited in his own country. He replied, that it was possible, but that he would shed the last drop of his blood to prevent it, and impose it as a sacred obligation upon his children to do the same. This was not said in vulgar bravado, but in unaffected earnestness. But it may be said that this is but one example. True, it is but one example, but I can assure the reader that it illustrates a universal sentiment.

It may be considered a little singular, but if the love of democracy admits of degrees in America, the ladies cherish it to the greatest extent. Could there be a better guarantee for its continuance?

Nor let it be supposed that the democratic sentiment is confined to the United States. The Canadas are now undergoing the probation to which the revolted colonies were subjected previously to 1776. At no period in their history were the provinces more loyal or well affected towards the mother country than they are at this moment. So long as they remain united with us, they will cherish as a sentiment the monarchical principle, albeit that their daily political practice is both of a republican character and tendency. But suppose a separation to take place, I candidly appeal to every Canadian to bear me out in the assertion that monarchy in Canada would be impossible. And whilst the tendency of things is thus towards democracy in our own colonies, some of us fancy that their tendency is towards monarchy in the Republic.

Many of what some regard as the more inflated

peculiarities of the American character, may be attributed to the faith which Americans cherish in the destiny of their country. Whatever may be its future social and political influence, they have no doubt that, as regards territorial extension, it will yet embrace the continent. The issues which such a consummation involves are enough to make a people feel proud of their country. The realization of their hopes in this respect, they regard as a mere question of time. They feel that there is, in reality, no power on the continent that can ultimately resist them. I was forcibly impressed with the extent to which this feeling prevails, on listening one day to a speech delivered by Mr. Crittenden of Kentucky, in the Senate, shortly after the breaking out of the Mexican war. It was in reply to Mr. Sevier from Arkansas, who was complaining that a portion of one of the counties of that State had been reserved to the Indians. Mr. Crittenden, in showing him how unworthy such a complaint was, reminded him that the whole State had been taken from the Indians, and not only it, but every State in the Confederacy. He then recapitulated the accessions made to the territory of the Union since the period of its independence. He alluded to the boundary, particularly the south and south-west, as ever changing, so as to embrace new acquisitions. It had first swept from the St. Mary's round the peninsula of Florida, and crept up the Gulf to the Mississippi and Sabine. It afterwards fled westward to the Nueces, and was then, he reminded the House, alluding to the cause of the war, supposed to be on the Rio Grande. It fled, he continued, before the Anglo-American race as it advanced. "Where is it now?" he asked in conclusion. "Just," he added, "where we please to put it."

Many fall into the mistake of supposing that an indulgence in hatred of England is a chronic state of the American mind. In the Irish population of the United States is the true source of the enmity towards this country which is sometimes exhibited. Originating amongst these, unscrupulous politicians fan the flame to serve their own purposes; but it has to be constantly supplied with fuel, or it speedily dies out. The feeling is not a general one, nor is it permanent with any section of the native population, not directly of Irish extraction. In all disputes with this country there is more of bluster than bad feeling. The American desires to see his country in advance of all nations, in power, wealth, and moral influence. Great Britain is the only power which he now regards as standing in the way. The Americans treat us as the only enemies, when enemies, worthy of a thought as such. It is this that makes them so touchy in all their quarrels with us. They are far more likely to be reasonable and conciliatory in a dispute with Spain than with Great Britain. They may give way in the one case, but they fear that if they did so in the other, it would seem as if they had been bullied into so doing. We, again, have been the only enemy with which they have ever been in serious collision. But after all, a friendly and kindly feeling with regard to us pervades the American mind; they would not willingly see us injured by a third party, if they could prevent it.

"We have had many quarrels with you," said a lady to me once in Washington, "but we are proud of our descent from the English! We court the French when it suits our purpose, but," she added, with great emphasis, "we would not be descended from them on any account."

The Americans are charged by some as being guilty of inconsistency in the fondness which they manifest for titles. But those who make this charge do so without reflection. The Americans are fond of titles, but that does not argue that they are inconsistent republicans. The fondness for titles which they display is but a manifestation of the fondness for distinction natural to the human mind. And what sane man ever inculcated the idea that republicanism was inconsistent with the love of distinction? Constitute society as you may, there must be posts of honour, power, influence, dignity, and emolument, to strive for. These exist in republics as well as under any other form of government. Are they not to be striven for without compromising one's political creed? And if the office is obtained, why not be called by its name? The Presidency of the republic is an office—he who obtains it is called the President. Does a man cease to be a republican because he aspires to both? Is it not rather a laudable ambition that prompts the aspiration? Or should he who obtains the office, drop the title? As it is with the title of President, so it is with all other titles in America. A judgeship is a distinction. On him who obtains it, it confers the appellation of Judge. A governorship of a State is a distinction. He who is appointed to it is called the Governor. And so on through all the offices in the State, civil and military. There is this broad and essential difference, however, between titles as coveted in America and titles as existing in Europe. There the title pertains to a distinction acquired by the individual himself, for himself, and has always connected with it some office of trust or responsibility. Here we have similar titles, but we have others also which spring from the mere accident of birth, which are connected with no duties, and which do not necessarily indicate any merit on the part of those possessing them. The time was in England when Marquis, Earl, and Viscount indicated something more than mere arbitrary social rank. There are in America no titles analogous to these. There duties are inseparable from titles. So long as there are offices in the Republic to be filled, and so long as republicans may legitimately aspire to fill them, so long may they, without sacrificing their consistency, assume the titles of the offices to which they are appointed.

The love of money is regarded by many as a striking trait in the American character. I fear that this is a weakness to which humanity must universally plead guilty. But it is quite true that it is an absorbing passion with the Americans. This cannot be denied, but it may be explained. America is a country in which fortunes have yet to be made. Wealth gives great distinction, and wealth is, more or less, within the grasp of all. Hence the universal scramble. All cannot be made wealthy, but all have a chance of securing a prize. This stimulates to the race, and hence the eagerness of the competition. In this country, however, the lottery is long since over, and with few exceptions the great prizes are already drawn. To the great bulk of the people wealth is utterly unattainable. All they can hope for is competency, and numbers fall short even of that. Men soon flag in a hopeless pursuit. Hence it is that, in this country, the scramble is neither so fierce nor universal.

Q 2

The American people discover an extraordinary talent for invention. The Patent-office in Washington is a most creditable monument to their inventive powers. They are also quick in the adoption of an improvement, no matter from what source it proceeds.

They are excessively fond of being well dressed. The artisans amongst them are particularly so, not so much from personal vanity, as from the fact that they make dress a test of respectability. Almost every man who is not an emigrant wears superfine broad-cloth in America, if we except the hard-working farmer, who generally attires himself in homespun. You seldom meet with a fustian jacket, except on an emigrant's back, in an American town.

This leads me, in concluding this chapter, briefly to glance at the physical condition of society America. If the social structure in the Republic has no florid Corinthian capital rising into the clear air above, neither has it a pedestal in the mire beneath. If it is devoid of much of the ornamental, so is it also wanting in much of the painful and degrading. It may not be so picturesque as many of the social fabrics which have sprung from chivalry and feudalism, but it is nevertheless compact, elegant, symmetrical, and commodious. It is to English society, what a modern house is to an Elizabethan mansion—it is not built so much to attract the eye as to accommodate the inmates.

The most important feature of American society, in connexion with its physical condition, is that competence is the lot of all. No matter to what this is attributable, whether to the extent and resources of the country, or to the nature of its institutions, or to both, such is the case, and one has not to be long in

America to discover it. It is extremely seldom that the willing hand in America is in want of employment, whilst the hard-working man has not only a competency on which to live, but, if frugal, may soon save up sufficient to procure for himself in the West a position of still greater comfort and independence. There are paupers in America, but, fortunately, they are very few. They are generally confined to the large towns; nor need they subsist upon charity, if they had the energy to go into the rural districts and seek employment. This, however, is not applicable to the majority of them, who are aged and infirm. It may be laid down as a general rule, without qualification, that none are deprived of competency in America except such as are negligent, idle, or grossly improvident. The general effect of this upon society has been already considered. Both in their social and political relations, all classes are thus able to act an independent part—an important consideration in connexion with the peculiar polity of America.

This being the broad and wholesome basis on which society, so far as regards its physical condition, rests, the character of the superstructure may easily be inferred. Where all classes have a competency, no class demurs to the luxuries enjoyed by another. There is but little jealousy of wealth in America, for reasons already explained. It is but in extremely rare instances that gigantic accumulations have as yet been made. Nor are they likely to be speedily multiplied, the whole spirit of legislation being against them. There is no legislation against accumulations of personal property, for the very good reason that it would be difficult to prevent its distribution. It is sure to circulate through the community, so that all, by turns, can have the advantage of it.   But the whole spirit of American legislation is decidedly averse to accumulations of landed property.   Such the people conceive would be incompatible with the safety of their institutions.   They have accordingly removed all restrictions upon its alienation, and land is now as marketable a commodity as the wheat that is raised upon it.

It is seldom indeed that you find a native American, or the descendant of an emigrant, occupying a lower position than that of an artisan.   Those who are mere labourers are almost exclusively emigrants, and in nineteen cases out of twenty, Irish emigrants.   Such as emigrate from England, Scotland, or Germany, are soon absorbed into the rural population, and become, by-and-by, proprietors of land themselves. But the Irish congregate in masses in the large towns, as they do here, to do the drudgery of the community.   It is thus that, if a canal is being dug, or a railway constructed, you meet with gangs of labourers almost entirely composed of Irishmen. Their descendants, however, become ambitious and thrifty, and form the best of citizens.

Enough has here been said to show that America is the country for the industrious and hard-working man.

CHAPTER XII.

A PEEP INTO THE FUTURE.

AN attempt at an hurried glance into the future, may
form a not inappropriate conclusion to the foregoing
general view of men and things in America down to
the present time. In turning the veil slightly aside,
we cannot expect to acquaint ourselves with the
details of destiny, whether of individuals or nations ;
but we may form some estimate of the more prominent of coming events from the palpable, albeit
undefined, shadows which they forecast. In attempting to fathom the future of America, we are lost
amid the multiplicity of speculations which crowd
upon us ; but we can, nevertheless, discern amongst
them some of the great purposes of fate, as they

loom upon us through the uncertain light, in obscure
outline but gigantic proportions.   Whatever may be
the fate of its present political arrangement, the
future of the Anglo-American Commonwealth is
pregnant with mighty destinies.

That into which we are first naturally led to
inquire, is the political future.   It is impossible to
foresee the changes which may be wrought in habits,
tastes, and opinions, during the flight of many successive generations; but, from what has been said in
the foregoing chapter, it will be evident, I think, that
for a long time at least, democracy, as the elementary principle of government in America, is sure to
maintain itself.   How rapidly or how frequently
soever systems may change, and others succeed them,
they will differ from each other in their form but
not in their substance.   Any form of government
but that which is essentially popular, is at present
impossible in America; and so far from things, as we
can now judge of them, tending towards monarchy,
they incline rather to the further extension of the
purely democratic element in the government.
Many point exultingly to what others again regard
despondingly, in proof that the tendency of things is
decidedly and rapidly towards monarchy—the prone-
ness which Americans exhibit to invest the successful
warrior with power.   It cannot be denied but that
this is an indefensible weakness in the American
character.   The accomplished and experienced statesman is frequently laid aside for the lucky or adroit
fighter; and men utterly untried in the important art
of administration, are suddenly cast by the wave of
popular enthusiasm into administrative positions,
because they have successfully conducted a campaign.

The art of administration, like that of war, is one which can only be acquired by experience. It does not follow that he who excels in one is necessarily prepared at once to grapple with the other. When there are tried generals at command, who would think of entrusting an important military expedition to him who had only approved himself as an accomplished statesman? But such a manifestation of confidence would not necessarily be more absurd than to put implicit faith in the administrative powers of a successful warrior, whose duties of administration have hitherto been confined within the precincts of the camp. We laugh at the idea of Lord John Russell taking the command of the Channel fleet, yet somehow or other we do not think it so very strange that Zachary Taylor should mount the presidential chair at Washington. But if Lord John Russell's antecedents have not prepared him for commanding the fleet, neither have General Taylor's prepared him for administering the civil government of the Republic. He may turn out to be a good President, but when the post to be filled was the highest civil post in the nation, to pass by such a man as Henry Clay, to promote General Taylor, was as inconsistent, on the part of the American people, as it would be on that of the government of this country, during a great national emergency, to supersede Admiral Sir W. Parker in the Mediterranean by the noble lord already named. Taylor, however, is not the first of the military Presidents. It was but a common act of gratitude to elevate General Washington to the presidency; in addition to which his powers of administration were great. Six civil Presidents succeeded him, after whom came

General Jackson, the very type of military Presidents. A civilian succeeded him, who was defeated in his second candidature by General Harrison. He again was followed by a civilian, who is about to be displaced by the hero of some recent victories. But both those who exult and those who despond at this hero-worship overrate its strength and misconceive its tendencies. The mistake is in believing that the hero, when elevated to power, might retain it. Sometimes, as was General Jackson's case, the idol is worshipped to excess; but the American people never lose sight of the fact that the idol is one of their own fashioning. Try to force one upon them, or let him be self-imposed, and see how long he would have a votary in the country. Had Jackson in the plenitude of his power manifested in the slightest degree an intention, or even a desire to perpetuate it, his most violent partizans would have seceded from him in a day. The Americans may make a man virtually dictator for a term of years, and obey him as such, but there is a limit in point of time to his sway which they will not permit him to transcend, and which the American executive is in itself powerless to extend. At the end of eight years General Jackson, like Cincinnatus, returned to his plough. Nobody wondered at it, because nobody was prepared for anything else. The periodic expiration of power in America is a law of its normal condition. Hero-worship in America, therefore, is not inconsistent with fidelity to the Republic, or with the continuance of that deep-rooted aversion to monarchy which pervades the American mind.

It is not, then, as to the duration of democracy in America that we need entertain any doubts, but as to the stability of the existing political arrangement.

It is not the republicanism, but the federalism of America that is in danger. By this I mean federalism in its present form and manifestation. For federalism and republicanism will co-exist there, although the present federal structure may be swept away. The only question, then, is as to the stability of the present Union. Will the American republics remain long united together as at present, or will they adopt a new form of political existence as one or as several confederations?

I have already glanced at the dangers as well as at the guarantees of the Union. The former chiefly resolve themselves into a conflict of material interests. The latter comprise strong material ties; some of a natural, and others of an artificial kind. Some sections of it sacrifice much in this respect to the Union. This sacrifice, to be continued, must be at least counterbalanced by the advantages and conveniences of the Union. The moment this ceases to be the case, disintegration would speedily ensue, but for the existence of other elements of cohesion. I allude to the national sentiment which pervades the American mind, and the national substratum on which the federal superstructure is based. But all these combined may not be proof against disturbing causes of a very violent description. Is the Union threatened with such at present?

It frequently happens that the greatest catastrophes are those which are the least heralded. A portentous calm sometimes precedes the earthquake, and the elements are often in the most perfect repose just as they are about to be most violently disturbed. It is true that there have been times when dissensions have exhibited themselves more angrily and more

noisily than at present.   But the Republic has never
yet struggled through a crisis like that which is
approaching it.   For the time being men's minds are
partly led away by other events of an interesting and
startling character, so that the premonitory symptoms
of the crisis are but partly heeded.   The *éclat* of a
successful war has not yet subsided, whilst the public
mind is still excited by the unexpected possession of
an El Dorado.   But despite of this, the difficulty
steadily approaches, unperceived and unheeded by
many, but increasing in magnitude every hour. Sooner
or later it would inevitably have presented itself, but
the Mexican war has, in its results, both precipitated
and aggravated it.   Slavery is the difficulty.   It is
the Ireland of the Americans.   A great question has
to be settled respecting it.   Its decision has hitherto
been from time to time postponed, from an instinctive
dread of its consequences.   The time of its solution
is now at hand.

The acquisition of so much new territory in the south-
west, whilst it has added to the national resources
and pandered to the national pride, has alarmed all
parties in connexion with its necessary bearing upon
the question of slavery.   From the moment in which
slavery is extended over it, the evil as regards the
continent is aggravated tenfold.   It is on this account
that the North is alarmed at the very thought of its
further extension.   From the moment in which the
territory is declared free, the South is placed in a
position of imminent peril.   Its property, its institutions, and the very existence of society in it, are
put in jeopardy.   A compromise is once more proposed, but the North is no longer disposed to stave
off an evil which must ultimately be grappled with.

By the adoption of the compromise, a large proportion of the acquired territory would be declared free ; but the North refuses to listen to it, and for very obvious reasons. Were it accepted, the line dividing the free from the slave regions would run across the continent to the Pacific. In other words, it would cover the whole of what remains of Mexico. Now there are few Americans who dream but that in the course of a very short time, another slice of Mexico will fall into their hands, and then another, and another still, until there is nothing left of the helpless Spanish republic. With these acquisitions in prospect, it would be impolitic in the extreme in the North to permit a broad belt of slave territory now to intervene between the free territory of the Republic west of the Mississippi, and the yet unappropriated provinces of Mexico. Should this be permitted, and some of these provinces be afterwards added to the Confederacy, the North could not well insist upon their being placed in the category of the free States. This is what now so greatly complicates the question of a compromise, even were the North still disposed, which it seems not to be, to stave off, for another period, the final decision.

The two sections of the Union have thus come at last, as it were, to a dead lock in reference to the question of slavery. It is important to the interests of each to carry its point ; it would be destructive to the policy of either to miss it. In other words, the time for drawn battles is past, and the period is approaching when one of the two sections of the Union must obtain, in connexion with this subject, a final and decisive victory over the other, or when the Union itself will be rent asunder. It is essential to the maintenance of the Union that one party or the other gives way. Will either do so? If so——which?

Both parties have already manifested their determination to oppose every resistance to the demands of the other. Since the meeting of Congress in December last, the North has been the aggressive party. The strongest exhibition which it has made of the spirit which animates it, has, as yet, been of an indirect kind, although intimately connected with the whole subject of slavery. I have already alluded to the important part which the District of Columbia plays in the whole question. It is essential to the interest of slavery that the institution in the District should be left intact. It exists in the District precisely as it exists in the circumjacent States of Maryland and Virginia; that is to say, not only are persons held to slavery in it, but they may also be trafficked in as slaves. The present House of Representatives has struck an incipient blow at the system in the District. It has, by a considerable majority on such a subject, adopted a resolution, ordering a bill to be introduced to prohibit in future all trafficking in slaves, in the District of Columbia. This gave rise to a most ominous excitement in Congress, and has created the utmost consternation throughout the length and breadth of the South. It is not that the interests of the slave States are bound up in the existence of a traffic in slaves in the District, but that they dread the slightest intermeddling with the subject on the part of Congress. Their object is to hold slavery in the District independent, in every respect, of Congressional action. They deny the power of Congress constitutionally to meddle with it in any degree. If it touches it in one point, it may touch it in all. The South, by sanctioning any proposal to legislate on the subject, would concede the whole question of power. But to this, as a vital point, it most tenaciously sticks. If it now permits Congress to abolish the traffic in slaves in the District, what is there to prevent Congress afterwards from abolishing slavery altogether in the District? This is the great object to which the North tends—it is the catastrophe which the South would ward off. It is but as a step towards it that the North seeks to introduce the bill alluded to—it is as a step towards it that the South resists its introduction. The North has, in other instances, also recently given token of the spirit which now animates it, but it is in connexion with this bill that it has assumed its most menacing attitude. It is high time that it took a final stand upon the subject. Slavery is admitted by all parties to be an evil which, more or less, affects the entire Republic. The North has all along submitted to it from the necessity of the case. It most unwisely aggravated it by the extension of slavery to Texas. It is now fully alive to the error which it then committed, and is not disposed to repeat it, for its repetition would be accompanied with the most formidable risks. Its tactic now is aggressive. Perhaps it would be wise, in the North, under all the circumstances of the case, to let the District alone, and to confine itself to resistance to the further extension of the system of slavery. But, not content with this, it is now attacking slavery in what is recognised as its citadel. The darling doctrine of the South, that it has no power to do so, involves this absurdity, that, if Congress has no power, there being no other legislative power in the District, slavery within the District is beyond all power. No State can touch it, and if Congress cannot do so, there is no power in the Union which can reach it.

The South lost no time in throwing itself into an attitude of determined resistance. By the last accounts, a species of committee of public safety was sitting on its behalf in the capital. When the resolution was adopted in the Lower House, a secession from that body of the southern members was proposed by a representative from the South. The proposal was cheered by some, and laughed at by others. But formidable passions have been roused, and Congress is treading upon a volcano. South Carolina is once more in a state of dangerous fermentation. Her leader and champion, Mr. Calhoun, the Slave King, is actively organizing resistance at Washington. The Southern members of both Houses had met under his auspices, to consider what was best to be done in the crisis. The result of their deliberations was the appointment of a committee to draw up an address to the South, pointing out to it its true position, real interests, and undoubted duty. The address was being drawn up, if not by Mr. Calhoun himself, at least under his directions. Speculation was rife as to its tenor and import. It was believed that it would openly advise the South that it had no longer anything to expect from the justice or forbearance of the North; and that the resistance, which it should offer to further aggression, should be influenced by this conviction. Should such be the scope and tenor of the address, the question is, how will the South receive it? There is danger in the way, whichever may be the mode in which it receives it. If warmly, the Southern members, supported by their constituents, will resist at all hazards. If coldly, the North will be stimulated to further encroachments, until the South is ultimately driven to the point of unanimous resistance.

Such is the crisis which has been superinduced by the spoliation of Mexico. California may yet cost more to the Union than all its gold can compensate for. Nations, as well as individuals, are amenable to the law of moral retribution.

It is not only in the new and perilous phase which it has given to the question of slavery, that the recent extension of its territory is fraught with danger to the Union. American politicians of the true Polk stamp are apt to trust too much to the capacity for expansion of the federal system. Hitherto it has safely expanded to admit territories which were not within its pale at the time of its foundation. But a power of extension does not necessarily imply a capacity for indefinite extension. Like the caout-chouc ring, the American system may contract so as to hold together only a few States, or it may expand so as to include many. But it should be remembered that, with every expansion, it becomes weaker and weaker, and that the strongest ligature will snap at last. The great danger, however, is not so much in the acquisition of new territories, as in the introduction of new interests into the Union. One of the main difficulties with which it has had to contend, was to reconcile the great interests which it included from the very first. It now embraces all that it could include, were it to absorb the continent. It embraced the manufacturing, the commercial, the agricultural, the cotton, and the sugar-growing interests, previously to the acquisition of California, which has comprehended within the catalogue that

connected with the precious metals.   So far as these
are concerned, therefore, its difficulties would now be
but little increased, were it to push its boundary to
the Isthmus.

The conflict of material interests has already
menaced the integrity of the Union.   And this, too,
when there were no other causes of irritation existing,
between section and section of the Confederacy.
That conflict is being renewed, and at a moment
when the public mind is agitated by other questions
of vast importance.   Unfortunately the question of
the tariff is one which the South regards, like that of
slavery, as sectional.   Notwithstanding the excellent
working of the tariff Act of 1846, the Whigs in the
Lower House have manifested a disposition, if they
could, to abrogate it.   Of course, so long as the
Senate is democratic, any attempt to revert to a high
tariff will prove abortive.   But it is this constant
attitude of defence, in which the South must keep
itself against the North, as well for the preservation
of its domestic institutions, as for the maintenance of
its material interests, that engenders that growing
feeling of alienation from the Union, which now to
some extent characterises the Southern mind.

Such are the difficulties which, by their combination, make up the present crisis.   If the Union
gets well through it, it may be regarded as indestruc-
tible.   If it splits upon the rock, what will be the
new political arrangements of the continent?

In that case, everything would seem to point to
the formation of two federal republics——the one in
the North and the other in the South——the one free,
the other slave holding.   The latter would strengthen
itself by engulfing Mexico——the former would gra-

dually absorb the Canadas. But natural though this division seems, a great difficulty lies in the way of its realization. That difficulty is the Mississippi. This river flows for half its course through free, for the other half through slave latitudes. Some of the States, which it binds together in one material and political system, are free, others slave-holding. With the exception of this difference, their interests are identical. Of course the slave States on the Mississippi would follow the fortunes of the slave States on the Atlantic and on the Gulf, whilst the free States on the Mississippi would make common cause with those on the lakes and on the sea-board. The result would be, that the Mississippi would then flow through two independent jurisdictions. Its lower half would be in possession of the Southern republic, without whose permission the States further up could make no use of it beyond the point separating the two jurisdictions. Would the States of the Upper Mississippi brook this partition of their common highway to the ocean? It is true that on account of the accessibility to them of the basin of the lakes and the St. Lawrence, and of the Atlantic seaports, by means of the artificial communications established between the valley and the coast, the Mississippi is less indispensable to them than to the States bordering it lower down. But it is, nevertheless, of the highest importance to them, and their reluctance to relinquish it would materially complicate the difficulties in the way of a new political arrangement.

A northern confederacy, embracing the north bank of the Ohio, and the whole basin of the St. Lawrence, would include the pith and enterprise of the continent. To such an arrangement I found many intelligent persons, both in New York and New England, looking forward, whilst the Canadians were gradually reconciling themselves to it. The divorce of the northern federation from the system of slavery would remove one very great objection which the Canadians entertain to the idea of a junction with the neighbouring States.

It is the extent to which we are interested in the material, that gives a passing interest to the political, future of America. Should a division of the Republic take place, there can be no doubt but that the closest commercial and political alliance would immediately spring up between the South and this country. Once free from the North, the South would reduce its tariff to the lowest revenue point, in order to promote the export of its great staple. The Southern market would in that case be more supplied than ever with fabrics by England, which would tend greatly to enhance the export of raw cotton to this country. It would be thus worth while to propitiate England ; for whilst the South would always be sure of the North as a market for her staple, she would not be so secure of England, who, if driven to it by interest or necessity, could procure her raw cotton elsewhere.

Whatever obscurity may now hang around the political future of the Republic, no doubt can exist as to the destiny of the different communities now constituting it, in a material point of view. There can be no question but that the material interests of the Union, as a whole, would be best subserved by the maintenance of its political integrity. Its disintegration would, however, have no very serious effect upon the development of the material wealth of the continent. And it is for this reason, that, in viewing America as our great industrial rival, we may pay but little regard to its political fortunes.

In estimating our own position amongst the nations of the earth, we are too little in the habit of taking the growing power, wealth, and influence of America into account. We think we do enough, when we measure ourselves against the nations of Europe, and take steps to maintain our supremacy amongst them. America is too far away to have much influence upon our political arrangements, and we accordingly attach but little consequence to her in any light. This is a great mistake. America is the only power on earth which we have to dread. We have not to fear her politically, for reasons already mentioned; we have not to apprehend any military chastisement at her hands, for in that respect we know both how to avenge and to defend ourselves; but we have to fear the colossal strides which she is taking in industrial development. We have less reason to dread the combined armaments of the world, than the silent and unostentatious operations of nature, and the progressive achievements of art, on the continent of America. We begird ourselves with fleets, and saturate the community with military and police, and think that we have done all that is needed for the perpetuation of our influence and the maintenance of our power. But in all this we mistake the real source of our power. What is it but our material wealth? Napoleon confessed that it was the gold more than the arms of England that humbled him. Our wealth is the result of our industry. It may be humiliating to confess it, but it is not by surrounding ourselves by all the pomp and panoply of war that we can maintain our position, but by the steady promotion and encouragement of our industry. Let our industry flag, and our unemployed capital will find investment elsewhere. Let capital once begin to flow from us, and the stream will soon become so broad and deep as to drain us, as a nation, of our life blood. Unless our industry is kept up, America will absorb our capital. It is like the magnetic mountain that extracted all the nails from the ship. Let us give it a wide berth or it will serve us in a similar manner, and leave us to sink with our cargo. We can only do this by—let me again repeat it—steadily and zealously promoting and encouraging our domestic industry.

I have already sufficiently explained the foundation which America has laid, both in the magnificent provisions of nature, and the stupendous achievements of art, for future material greatness. Her resources in almost every point of view are infinitely greater than any that we possess. Look at her forests, her fertile valleys, and vast alluvial plains. Look at the variety of her productions, including most of those that are tropical, and all that are yielded by the temperate zone; and look at her mines teeming with coal, iron, lead, copper, and, as has been just discovered, with silver and gold. Look again at her enormous territory, and at the advantages she possesses for turning all her resources into account, in her magnificent systems of lakes and rivers; in her extensive sea-coast; in her numerous and excellent harbours; and in her geographical position, presenting, as she does, a double front to the Old World, or holding out, as it were, one hand to Asia, and the other to Europe. But such resources and advantages are only valuable when properly turned to account. It is only by their being so that they will become formidable to us. We have only to look to the race possessing them to decide whether they are likely to be turned to account or not. The Americans are Englishmen exaggerated, if any thing, as regards enterprise. This is not to be wondered at, as they have, as a people, more incentives than we have to enterprise. Of this we may rest assured, that the most will be made of the resources and advantages at their disposal. This is all that has made us great. We have turned our coal and our iron, and our other resources, to account, and the world has by turns wondered at and envied the result. The American stock of coal and of iron is more than thirty times as great as ours, and more than twelve times as great as that of all Europe. Their other resources are in the same proportion, as compared with ours. And if our resources, turned to good account, have made us what we are, what will be the fabric of material greatness which will yet spring from the ample development of resources thirty times as great? If the industry of from twenty to thirty millions of people, with limited means, have raised England to her present pinnacle of greatness and glory, what will the industry of 150,000,000 yet effect in America, when brought to bear upon resources almost illimitable? The continent will yet be Anglo-Saxon from Panama to Hudson's Bay. What Anglo-Saxons have done, circumstanced as we have been, is but a faint type of what Anglo-Saxons will yet do, working in far greater numbers, on a far more favourable field of operation.

It is the consideration that America will yet exhibit, in magnified proportions, all that has tended to make England great, that leads one irresistibly, however reluctantly, to the conclusion that the power of England must yet succumb to that of her offspring. There is, however, this consolation left us, that the predominant influence in the world will still be in the hands of our own race. That influence will not pass to a different race, but simply to a different scene of action. It has been England's fate, during her bright career, to plant new States, which will inherit her power and her influence after her. On the continent of North America, on many points on the coast of South America, at the southern extremity of Africa, throughout wide Australia, in New Zealand, in Van Diemen's land, and the Indian Archipelago, the Anglo-Saxon race will prevail, and the Anglo-Saxon language be spoken, long after England's glories have become historic and traditional. These different communities, flourishing remote from each other, will all be animated by a kindred spirit, and will cherish a common sentiment of attachment to their common parent, who will long exercise a moral influence over them, after her political power has been eclipsed. Not that England will not always be able to maintain her position in Europe. The powers which are destined to overshadow her are springing up elsewhere, and are of her own planting. Of these the American Republic, or Republics, as the case may be, will both politically and commercially take the lead, when England, having fulfilled her glorious mission, shall have abdicated her supremacy, and the sceptre of empire shall have passed from her for ever.

A CHAPTER ON CALIFORNIA.

IT is related of Columbus, that during one of his voyages he coasted along the southern shore of Cuba, with a view to verify his own impression that it was an island. After sailing for many days to the westward, his men became mutinous and unmanageable, and he was compelled to put back when within half a day's sail of the western extremity of the island. Had he pursued his way for a few hours more, he would have taken a northward course, which would have brought him to the mouths of the Sabine, the Mississippi, the Mobile, and the Appalachicola. The effects which so simple an event might have had upon the destinies of the Continent, it is not now easy to speculate upon. The chances would have been, however, that the whole course of Spanish discovery and settlement would have taken a northerly direction, and that the America, which is now Anglo-Saxon, would have passed under the dominion of the crown of Spain, and been peopled by a Spanish race.

It is also related of Sir Francis Drake, that when cruising off the north-west coast of America, he landed in California, and traded with the natives. He was in search, as Raleigh had been before him, of golden regions in the West. He was at San Francisco, but never reached the Sacramento. Had he done so, and discovered the soil saturated with gold, how different a turn might have been given to the destinies of the Continent! It is by such simple events that the fortunes of nations and continents are sometimes most profoundly affected. Providence had better things in store for the continent of North America than would probably have fallen to its lot had Columbus doubled the western point of Cuba, or Drake discovered the buried treasures of the Sierra Nevada. It was some time afterwards ere the insular character of Cuba was known, and but a few months have as yet elapsed since the mineral value of California has been disclosed to the world.

The chapter which I am here induced to add respecting this latest acquisition of the Republic, has no necessary connexion with the preceding part of this work. My reason for adding it is partly to be found in the intrinsic interest of the subject, and partly in the effects which it is likely to produce on the future fortunes of the Republic. Heretofore I have described nothing but what I have seen. I make no pretensions to have seen California: but what follows of a descriptive character respecting it is not drawn from the numerous accounts of it which have recently been given to the public, some of them authentic and some of a surreptitious character, but from what I heard concerning it in the Senate of the United States, delivered by one who is intimately connected with that meritorious officer, Captain Fremont, who has done more than any other *employé* of the American government to extend our knowledge of Upper California.

It may be as well first to describe its geographical position and extent. It is the northern section of an enormous tract of country, resting on the Pacific

Ocean, and for many years forming a province of the Mexican republic, under the name of California. It was afterwards divided into two ; the peninsula of California forming the old or lower province, and the vast tract extending from the head of the gulf to the 42d parallel of north latitude, and from the Pacific to the Anahuac Mountains, being erected into a separate province, under the name of Upper, or New California. Its total length upon the Pacific is about 700 miles, and it varies in breadth from 600 to 800 miles. Taking 700 miles as its mean breadth, its area will be 490,000 square miles, being more than double the size of France, and nearly quadruple that of Great Britain. Between it and the States on the Mississippi extends a vast irreclaimable desert of nearly a thousand miles in width. It is thus a region more effectually separated from the populous portion of the Union, than if so much sea intervened between them. It has but few good harbours, but the Bay of San Francisco, the best of them, is one of the finest in the world.

Upper California is divided into two great sections, separated from each other by the Sierra Nevada—a chain of lofty hills, which pursues, throughout its whole length, a parallel course with the Pacific, from 150 to 200 miles back from the coast. The section lying between this mountain chain and the coast is by far the smaller of the two—the other, which lies to the eastward from the Sierra Nevada to the Rocky Mountains, comprising fully four-fifths of the whole area of California. Much obscurity hangs over the character and capabilities of this enormous tract of territory. That it is fertile in the immediate vicinity of the mountains which bound it on the east and on the west, there can be no doubt; whilst the natural capabilities of the portion of it which abuts towards the south-east upon New Mexico, are known to be as great as those of any other section of the continent. But the enormous area which passes under the general name of the Great Interior Basin of California, is as unknown to us as is Central Australia. It will not long remain so, however, the American government having already taken effective steps for its survey. This vast district has been skirted by various ex-plorators, but none have as yet had the courage or the means of penetrating into the interior. So far as it has been examined, it appears to present many features analogous to those which we know to cha-racterise, to some extent, the interior regions of the Australian continent. A little distance back from the hills, it becomes sandy and arid ; the streams seem to flow internally, and bodies of salt water have been discovered in it. To those familiar with the history of Australian exploration and discovery, this will recall many of the physical phenomena of that extraordinary region.

The coast section, lying to the westward of the Sierra Nevada, is better known, and in every way better adapted for the habitation of man. It extends in one elongated valley from the most northerly limit of the territory to the head of the Gulf of California. This valley is enclosed between the Sierra Nevada, and a range of low hills known as the Coast range, and lying but a short distance back from the Pacific ; indeed, at many points they dip sheer down into the ocean. This range, after traversing Upper Cali-fornia, pursues its way southerly through the penin-sula of California, of which, in fact, it forms the basis. The Sierra Nevada, diverging a little to the eastward, continues its southerly course, but under different names, through the Mexican province of Sonora. The Gulf of California here intervenes between them, as the valley does higher up; the gulf being, in fact, a continuation of the valley, but on so low a level that it is invaded by the Pacific. The valley thus extends from the head of the gulf to beyond the line dividing Oregon from California, and has a mean width of about 125 miles. This valley constitutes, so far as it has yet been discovered, the gold region of California.

Such being the geographical position, extent and configuration of California, it may be as well now to consider briefly the capabilities of its soil and the nature of its climate. Of the character of the great region lying to the east of the Sierra Nevada, but little that is authentic, as already intimated, is known. The inference, however, drawn by those most capable of judging is, that nearly two-thirds of it is a desert, the arid waste being surrounded by a belt of fertile land lying under the shelter of the Sierra Nevada on the west, and under that of the Rocky Mountains on the east. Towards the north this fertile belt rests on a chain of small lakes which lie near the Oregon line, whilst on the south it skirts the province of Sonora. This belt is capable of producing every species of grain raised within corresponding latitudes on the Atlantic side of the Continent.

But by far the most valuable portion of the territory in regard to soil is the valley already alluded to as constituting the coast region. The soil of the valley is in most places fertile to a degree, producing in abundance not only Indian corn, rye and barley, but also wheat, the

olive and the vine.   It is well irrigated by streams, few
of which descend from the Coast range.   From the
direction taken by its streams, the valley seems to
have three great inclinations : one descending towards
the head of the Gulf, a portion of the Colorada, the
largest river of California, passing through it ; another
descending northward towards the Bay of San Francisco, watered by the San Joachim and its tributaries ;
and the third dipping towards the south to the same
point, watered by the Sacramento and its tributaries.
The Colorado descends from the Rocky Mountains, not
far from where the Rio Grande and the Red River
take their rise to flow to the opposite side of the continent.   Both the San Joachim and the Sacramento
are almost exclusively formed by the numerous
streams which descend from the westerly slopes of the
Sierra Nevada.   These streams, which have but brief
courses, run almost parallel to each other, in the
direction of the Pacific, until they reach the lowest
level of the valley, when the land begins to rise again
to form the Coast range.   Here they find their way
by a common channel to the Bay of San Francisco,
the San Joachim flowing due north, and the Sacramento
due south to the bay.   The two main streams, by
which the different rivers descending from the Sierra
thus find their way to the ocean, flow, for almost
their entire course, parallel to the two ranges of
mountains which enclose the valley.   Both are much
nearer to the Coast range than they are to the Sierra.
It will thus be seen, that so far as irrigation is concerned, nature has done everything for this favoured
region.   With the exception of its more southerly
portion, which dips towards the Gulf, it is traversed
in its whole length by the two streams just named,

which are but the collections of the waters of the innumerable rivers which, having their rise in the Sierra, flow westward till they reach the bottom of the valley. The region thus drained into the Bay of San Francisco is about 500 miles long, and from 100 to 150 wide. The elongated basin constituting it appears at one time to have been covered with water, which at length so accumulated as to break its way through the Coast range to the Pacific at the point now forming the bay.

It would be but reasonable to infer, even had we no positive information upon the subject, that a district so well irrigated must be fertile. Such is the case with the coast region of California. Its agricultural capabilities attracted to it the attention of American settlers, before its incorporation with the Union was determined upon, and before its golden treasures were dreamt of. Its wealth, in an agricultural point of view, consists so far chiefly of live stock. Its exports of hides and tallow have been considerable. It has also traded very largely in furs.

The wheat produced in the fertile districts of California is of a very superior description, and the annual product is large, except in years when droughts are severe and protracted. Nor has California been backward in the produce of this staple, which it has exported in considerable quantities, both to Oregon and to Russian America. Peas and beans are also easily produced, whilst Indian corn flourishes as an indigenous grain. Grapes can not only be raised, but have been produced to a great extent, and considerable quantities of wine have been made from them. Cattle, sheep, mules, horses, goats and swine are abundant. The mutton of California is described

as of the best flavour, although the wool is very inferior, from the want of care in tending the sheep.

Rather unfavourable impressions have long prevailed as to the climate of California. That of the
peninsula, which for a long time was the only portion
of the territory at all known, is exceedingly dry, the
country being sterile, chiefly for want of rain. It
has been supposed that the same is the case as
regards the whole region. This is a mistake. With
the exception of occasional droughts, the coast section
of California is well supplied with rain; the clouds
produced by the evaporations of the Pacific being
deprived of their superabundant moisture by the Coast
range and the Sierra Nevada. In the peninsula the
hills are not high enough to arrest the clouds, which
float over it to fertilize the soils of Sonora and New
Mexico. In the snows which perennially crown the
Sierra in Upper California, its coast region has a
never-failing fountain for the supply of its streams.
What becomes of the rivers which descend the Sierra
on its eastern side and flow towards the interior, is
one of the most interesting of the problems which
have yet to be solved with respect to the great interior
basin. That the climate of that basin is much drier
than that of the coast region, is obvious from the
Sierra intercepting the clouds, which proceed from
the only quarter, the west, from which they there
bring rain. But in the heavy dews which fall, particularly in the vicinity of the Rocky Mountains,
nature has provided a species of compensation for the
want of rain. It is to these dews that many of the
most productive districts of New Mexico, which has
been incorporated along with California into the
Union, owe their fertility.

Such, then, is the region which the late war with Mexico has added to the territories of the United States. It presents a broad fertile belt upon the Pacific, a sweep of productive territory, extending around the interior basin, and the exuberant province of New Mexico, rich both in agricultural and in mineral wealth. The new acquisition was considered a great prize before the Valley of the Sacramento disclosed its hidden treasures. And so it was, for it would not be easy to over-estimate its importance to the Union in a commercial or a political light. The province of New Mexico forms its south-easterly portion. In more points than one is this portion of it an important acquisition. From the Lakes down to New Mexico a vast desert intervenes, as already intimated, between the States on the Mississippi and the territory on the Pacific, from which new States will yet spring. It is not until we descend to the latitude of New Mexico that we find the continent crossed from sea to sea by a tract of fertile and practicable country. This province will thus form an important link in the chain of communication which will yet be established between the two sea-boards. In addition to this, it will, from its known mineral wealth, and the fertile character of the numerous valleys by which it is intersected, soon attract to it a large and enterprising population. The importance of having such a population midway between the two sides of the continent is obvious. They will yet constitute the hardiest of the heterogeneous population of the Union, the country which they will inhabit being of a rugged and mountainous character. Indeed, New Mexico and the south-eastern portion of California may be regarded as the Switzerland of America.

It did not require, therefore, the recent discovery in the valley of the Sacramento to make their new acquisition valuable in the eyes of the American people. That event has not only enhanced its value to them, but has attracted to it universal attention.

In a former part of this work, whilst traversing with the reader the Southern Atlantic States, I drew his attention to the only region in the Union then known as the gold region. I described it as extending from the basin of the St. Lawrence in a south-westerly direction to the northern counties of Alabama. The length of this region is 700 miles, and its average width is from 80 to 100. In approaching Alabama, it diverges into Tennessee. It lies chiefly to the east of the Allegany Mountains, and between their different ridges. Some branches of it have been traced west of the mountains. Throughout the whole of this region gold is found in more or less quantity, the auriferous belt being richest in its yield in North Carolina and Virginia. But, as already shown, it has not proved itself sufficiently productive at any one point to be very extensively or systematically worked. The gold is generally found in the beds of the rivers or by their banks, the great bulk of that produced having been so by washing it from the deposit in which it is found. In some instances it has been found in lumps, embedded in slate and quartz. When I was thus describing this auriferous belt lying at the bases of the Alleganies, the gold region of California was unknown to Europe. From the descriptions which we have since received of it, both in connexion with its geological formation and the state in which the gold is found in it, it appears to present many points of analogy to the gold region on the Atlantic side of

the continent. So far as that of California has yet
been discovered, it is nearly equal in extent to the
other, its length being 600 miles, and its width over
100. The two regions differ more in the quantity of
gold which they yield than in its quality, or in any
other circumstance with which we are acquainted
connected with them in their auriferous capacity.

There can be little doubt but that the origin of the
gold found in the valleys of the San Joachim and the
Sacramento, is the Sierra Nevada. It has for ages
been washed down into the plain by the torrents
descending from the mountains. That the whole
range is rich in the precious ore is evident from the
extent to which it has been found in the valleys, and
the quantities in which it has been discovered in the
rocks and amongst the hills. Whether mines will
yet be opened in the mountains and worked, it is very
difficult to say. The rich treasures which they enclose may lie beyond the line of perpetual congela-
tion, where they will bid defiance to the approaches
of man. It is by no means improbable that the great
interior basin is skirted on the west by an auriferous
belt, for the golden torrents may have flowed down
both slopes of the Sierra.

Many are prone to believe that the gold of California is only to be found on the surface, and that its
stock will soon be exhausted. The state in which it
is discovered in the valley, is no criterion of the nature
or productiveness of the mines in the mountains. So
far as the gold has been discovered, not in the position to which it has been washed by successive torrents, but imbedded in the rock at the bases of the
Sierra, it certainly comes very near the surface. But
if we are guided by the analogy afforded by almost

all the American mines now worked, this does not make against the productiveness of the gold mines of California. Almost all the mineral wealth of the Union, hitherto discovered, develops itself close to the surface. In some cases the coal of Pennsylvania is mixed with the very soil; whilst, at some points, the great coal-bed of Virginia approaches within a few feet of the surface. The iron ore in most of the States is also found at but little depth. The lead in the north-western section of Illinois lay in such quantities on the surface, that the Indians, who had no notion of mining, used to turn it to account. And so with the copper in the vicinity of Lake Superior—huge masses of it being sometimes found lying exposed to the sun. Yet, notwithstanding their superficial richness, all these mines are found to be productive to a great depth, whilst in many cases the deeper they are worked the more productive do they become. Judging, therefore, from what is known of the disposition and extent of the mineral wealth of the continent, from the Lakes to the Gulf, and from the Atlantic to the Mississippi, no inference need be drawn of the poverty of the mines of California, from the gold being found either upon, or close to the surface.

There are several routes from the Atlantic seaboard to California, but the safest and most practicable at present is that by Panama. From that city to the Columbia a line of steamers has been established, each steamer calling on its way north and south, at San Francisco or Monterey. Parties not choosing to proceed by this route, may cross the desert from Missouri, and descend upon the Pacific, after penetrating the defiles of the Rocky Mountains, and those of the Sierra Nevada; but it is necessary in taking this route to proceed in great numbers, in fact to form a caravan, such as is formed to cross the deserts of Africa. There is another route by Santa Fé, through New Mexico. This will, undoubtedly, at no distant day, be the main route to the Pacific. The sea voyage round Cape Horn is from 15,000 to 17,000 miles in length; a voyage which few parties will undertake, but such as may be driven by necessity to do so.

Twenty years will not elapse ere the Atlantic and Pacific are connected together by a line of railway. The construction of a railway from the Mississippi to the mouth of the Columbia was seriously spoken of in 1846, and, during my stay in Washington, more than one plan for such a project was presented to Congress. This was before the Republic had added 700 miles of coast to its territory on the Pacific, and consequently before the gold region of California became the property of the Union. If a railway was talked of as a desirable thing then, its construction is likely to be expedited now.

It is impossible at present to calculate the effect which this startling discovery is likely to have upon the destinies of the Union. If gold abounds in California to anything like the extent supposed, the consequences will be such as to embrace the whole civilized world. The bullion market will be seriously affected, and gold will become abundant as a medium of exchange. This will be a most desirable result to see accomplished. But there is another point of view in which the discovery will be attended with the most important consequences. Hitherto the Pacific side of America has played but an insignificant part in the commercial and political arrangements of the world. Emigrants are now flocking to it from all quarters ; and many years will not elapse ere numerous and energetic communities extend from Vancouver's Island to the head of the Gulf of California. These communities will not only traffic with South America, but they will also institute a trade with Asia. Means of speedy personal transit between Asia and America will soon follow, and the shortest route from Europe to Canton will yet be by the Bay of San Francisco. When the circumstances exist which will give rise to these arrangements, how far they may revolutionize the interests of the world it is now impossible to tell. It is evident, however, that the time is near at hand when the Asiatic trade of America will be carried on across the continent, and when the United States will form, as it were, the stepping-stone between Western Europe and Eastern Asia. This will complete the political and commercial triumph of America.*

* As stated in a note to a former part of the work, the subject of a great line of railway across the continent has been earnestly taken up by some of the leading statesmen of the Union, and neither its inception nor its completion can now long be delayed.

So far the gold regions of California have, on the whole, not falsified the expectations that were formed of them.

So great has been the flow of population to the Pacific Coast, that even already is California prepared for admission as a State into the Union. Its admission will take place during the present session of Congress, unless the attitude which it has assumed, in reference to the subject of slavery, should induce the South to throw difficulties in the way. Should there be sufficient to delay the event, California may not again ask for admission, but declare itself independent. In that case, it would be a greater thorn in the side of the South, than as a free member of the Confederacy.

THE END.

LONDON :
R. CLAY, PRINTER, BREAD STREET HILL.

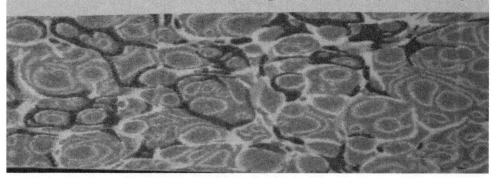

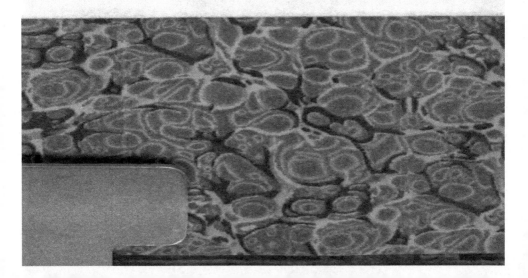